Science
and Subjectivity

Second Edition 1982

Israel Scheffler

Hackett Publishing Company

BD220
S2
1982

Hackett Publishing Company
Box 44937
Indianapolis, Indiana 46204

Second Edition
Second Printing 1985

Library of Congress Catalog Card No.: 81-85414

ISBN 0-915145-30-8 (pbk)
ISBN 0-914145-31-6

Contents

Foreword to the Second Edition

It is a pleasure for me to introduce this new edition of *Science and Subjectivity*. Although almost a decade and a half have elapsed since the book first appeared, I believe that the situation it addressed is still urgent, and that the treatment it presented remains appropriate. Were I writing the book today, I should alter some details and elaborate various passages; I should certainly, however, retain its fundamental characterization of the problems and uphold its major criticisms and lines of argument.

The main purpose of the book is the reinterpretation and defense of the ideal of objectivity in the light of recent criticisms that have brought it under severe attack. Since without objectivity as a guiding ideal there can be no science or, indeed, any rational deliberation whatever, the task undertaken by the book is pressing and far-reaching in its ramifications. Whether my philosophical effort has been successful or not is for others to judge. I hope, above all, to have brought home the seriousness of the issue; let those unpersuaded by my response join in the inquiry and do better.

I would hope, in any case, that future readers of my book might avoid the most frequent mistakes and misinterpretations evident in discussions of it to date. To begin with, my elaboration of the "standard view" of science in Chapter 1 is intended as an *initial basis* for discussion, helpful in explaining recent criticisms of objectivity; it is not put forward as a *thesis* I defend. The view of objectivity I do espouse emerges from an analytical scrutiny *both* of the "standard view" and of the several criticisms in question. Though I defend against these criticisms of the objective ideal, I uphold, not the "standard view" of this ideal, but rather a subtler version incorporating whatever of truth I find in the criticisms themselves. Thus, for example, I describe the "standard view" as holding that observation supplies "hard phenomenal data" for hypotheses to encompass; in Chapter 2, on the other hand, I argue for "the continuity and interaction of observation and conceptualization," rejecting the doctrine of pure, ineffable, or infallible phenomenal data. In sum, I do not simply defend the ideal of objectivity as traditionally formulated; in the course of defense, I reinterpret it.

As it is a mistake to think I affirm the fixity of observation, it is equally an error to take me as upholding the fixity of meaning. What I rebut, in Chapter 3, is the view that alterations of theoretical framework inevitably alter descriptive meanings. The negation of this view, which I assert, is that alterations of theoretical framework are compatible with the constancy of descriptive meanings. But the latter proposition clearly does not imply that such meanings are inevitably or always constant. Meaning-identity "may, of course [as I say in Chapter 3], be deliberately changed for independent reasons. But the mere fact of absorption into varying frameworks of theory does not, in itself, require us to say that the old laws have altogether changed, giving way to new." In short, that *meaning change* is not inevitable does not show that *meaning constancy* is inevitable; I reject the inevitability of the one, and also of the other.

By far the widest notice received by *Science and Subjectivity* has been directed to Chapter 4, largely devoted to a discussion of Thomas Kuhn's *The Structure of Scientific Revolutions*. Since I have replied to certain criticisms of that chapter, summarizing and also supplementing its basic arguments in my 1972 paper "Vision and Revolution: A Postscript on Kuhn," I am very pleased that the present edition makes that paper available as *Appendix A* below.

The wide attention given to Chapter 4, welcome as it may be, has tended to overshadow the discussion of the following chapter, treating the opposition of *coherence* and *certainty*. I should therefore emphasize that the view of objectivity worked out in earlier portions of the book is incomplete without a consideration of what Schlick refers to as the "contact between knowledge and reality," discussed in connection with his debate with Neurath, in Chapter 5.

My conception of objectivity rests at no point upon a sharp division of cognition from emotion. I have insisted, in Chapter 1, that "scientific habits of mind are compatible with passionate advocacy, strong faith, intuitive conjecture, and imaginative speculation." Moreover, Chapter 2 argues—from the existence of *surprise*—that what we expect does not wholly determine what we see, the disharmony constituting a basic source of observational control over belief. Prompted by recurrent questions by students regarding the role of surprise, and reflecting further on Schlick's references to the "joy of verification" (discussed in Chapter 5), I attempted a general statement as to how cognition incorporates emotional elements in its own functioning. The result was a paper that treated the organization of emotions in inquiry, and then introduced a concept of *cognitive emotions,* of which surprise served as a chief example. This paper, "In Praise of the Cognitive Emotions," was first presented in 1976 and published in 1977. It provides a substantial elaboration of my earlier treatment in the book and I am glad that it appears in the present edition as *Appendix B*.

The general discussion of themes relevant to the concerns of *Science and Subjectivity* has been too extensive since 1967 to allow for summary or commentary in this place. I should like only to call attention here to some contributions that have appeared during this period, congenial to the point of view expressed by the book, and emphasizing or supplementing certain of the arguments put forward therein. (Of course, it is not implied that there is full agreement between the authors mentioned and myself on the issues

x   SCIENCE AND SUBJECTIVITY

treated.) Michael Martin, in two papers, extended certain of the considerations of *Science and Subjectivity,* presenting arguments to show that not only is sense invariance unnecessary, but even referential invariance is unnecessary for objective theoretical comparisons (See Martin, "Referential Variance and Scientific Objectivity," *British Journal for the Philosophy of Science,* vol. 22 (1971), 17–26, and "Ontological Variance and Scientific Objectivity," Ibid., vol. 23 (1972), 252–256.) C. R. Kordig, in his book, *The Justification of Scientific Change* (1971), presents clearly a point of view similar in various respects to my own. And Harvey Siegel, in his paper, "Meiland on Scheffler, Kuhn, and Objectivity in Science," *Philosophy of Science,* vol. 43 (1976), 441–448, responds to criticisms of some of my remarks on Kuhn in J. W. Meiland, "Kuhn, Scheffler, and Objectivity in Science," *Philosophy of Science,* vol. 41 (1974), 179–187.

Newton, Massachusetts   ISRAEL SCHEFFLER
October, 1981

Preface

That the ideal of objectivity has been fundamental to science is beyond question. The philosophical task is to assess and interpret this ideal: to ask how, if at all, objectivity is possible. This task is especially urgent now, when received opinions as to the sources of objectivity in science are increasingly under attack. The notion of a fixed observational given, of a constant descriptive language, of a shared methodology of investigation, of a rational community advancing its knowledge of the real world—all have been subjected to severe and mounting criticism from a variety of directions.

The overall tendency of such criticism has been to call into question the very conception of scientific thought as a responsible enterprise of reasonable men. The extreme alternative that threatens is the view that theory is not controlled by data, but that data are manufactured by theory; that rival hypotheses cannot be rationally evaluated, there being no neutral court of observational appeal nor any shared stock of meanings; that scientific change is a product not of evidential appraisal and logical judgment, but of intuition, persuasion, and conversion; that reality does not constrain the thought of the scientist but is rather itself a projection of that thought. Unless the concept of responsible scientific endeavor is to be given up as a huge illusion, the challenge of this alternative must, clearly, be met. A philosophical examination, not only of current criticisms, but of the epistemological bases of a viable objectivism, is thus a matter of first importance. In the following lectures, I outline some efforts to contribute to such an examination.

I am grateful to Oberlin College for having invited me to deliver the Mead-Swing Lectures in May of 1965. These lectures, on the theme of "Science and Subjectivity," constituted the basis of the first four lectures in this book. My sabbatical year of 1965–66 was spent at the Center for Cognitive Studies at Harvard, and was devoted largely to revision and elaboration of my original drafts. The result is the present set of five lectures, the first four considerably expanded and the last newly written. During 1965–66, I received support from the National Science Foundation (Grant No. GS–644), for which I am very grateful. To the Center, I am indebted for its friendly hospitality throughout my stay.

I want to thank Harold C. Weisberg and Marx W. Wartofsky for having provided me with instructive criticisms of my manuscript. To Henry D. Aiken, Noam Chomsky, Nelson Goodman, Sidney Morgenbesser, W. V. Quine, and to Weisberg, I am generally indebted for illuminating discussions of various related topics. Lester V. Hoffman worked as my research assistant in 1965–66, providing me not only with bibliographical aid but with substantive criticism and stimulating commentary. And I am grateful to Mrs. Dorothy Spotts for her secretarial assistance and her typing of several versions of the manuscript.

Newton, Massachusetts   ISRAEL SCHEFFLER
November, 1966

1

Objectivity Under Attack

A FUNDAMENTAL FEATURE of science is its ideal of objectivity, an ideal that subjects all scientific statements to the test of independent and impartial criteria, recognizing no authority of persons in the realm of cognition. The claimant to scientific knowledge is responsible for what he says, acknowledging the relevance of considerations beyond his wish or advocacy to the judgment of his assertions. In assertion he is not simply expressing himself but making a claim; he is trying to meet independent standards, to satisfy factual requirements whose fulfillment cannot be guaranteed in advance.

To propound one's beliefs in a scientific spirit is to acknowledge that they may turn out wrong under continued examination, that they may fail to sustain themselves critically in an enlarged experience. It is, in effect, to conceive one's self of the here and now as linked through potential converse with a community of others, whose differences of location or opinion yet allow a common discourse and access to a shared world. It is accordingly to lay oneself open to criticism from any quarter and to acquire an impersonal regard for the judgments of others; for what matters is not

1

who they are, but whether they properly voice the import of controlling standards. Assertions that purport to be scientific are, in sum, held subject to control by reference to independent checks.

Commitment to fair controls over assertion is the basis of the scientific attitude of impartiality and detachment; indeed, one might say that it constitutes this attitude. For impartiality and detachment are not to be thought of as substantive qualities of the scientist's personality or the style of his thought; scientists are as variegated in these respects as any other group of people. Scientific habits of mind are compatible with passionate advocacy, strong faith, intuitive conjecture, and imaginative speculation. What is central is the acknowledgment of general controls to which one's dearest beliefs are ultimately subject. These controls, embodied in and transmitted by the institutions of science, represent the fundamental rules of its game. To devise fair controls for new ranges of assertion, and to guarantee the fairness of existing controls in the old, constitute the rationale of these rules. The cold and aloof scientist is, then, a myth.

It must be emphasized that the function of scientific controls is to channel critique and facilitate evaluation rather than to generate discoveries by routine. Control provides, in short, no mechanical substitute for ideas; there *are* no substitutes for ideas. The late Hans Reichenbach drew a sharp distinction in his philosophy of science between the "context of discovery" and the "context of justification,"[1] and he was right to do so. For the mechanical scientist is also a myth.

Now, the ideal of objectivity, as thus far described, characterizes not only the scientist, but also the historian, the philosopher, the mathematician, the man of affairs—insofar as all make cognitive claims in a rational spirit. A parallel ideal is relevant for the moral person as well. The ideal of objectivity is, indeed, closely tied to the general notion of rationality, which is theoretically applicable to both the cognitive and the moral spheres. In both spheres, we honor demands for relevant reasons and acknowledge control by principle. In both, we suppose a commitment to general rules capable of run-

[1] Hans Reichenbach, *Experience and Prediction* (Chicago: The University of Chicago Press, 1938), chapter 1, sec. 1.

ning against one's own wishes in any particular case. In neither sphere is personal authority decisive; as S. I. Benn and R. S. Peters have put it,

> The procedural rules of science lay it down . . . that hypotheses must be decided on by looking at the evidence, not by appealing to a man. There are also, and can be, no rules to decide who will be the originators of scientific theories.
>
> In a similar way . . . a rule cannot be a moral one if it is to be accepted just because someone has laid it down or made a decision between competing alternatives. Reasons must be given for it, not originators or umpires produced. Of course, in both enterprises provisional authorities can be consulted. But there are usually good reasons for this choice and their pronouncements are never to be regarded as final just because they have made them. In science and morality there are no appointed judges or policemen.[2]

There is thus no ground for restricting the applicability of the *ideal* of objectivity to *de facto* science, as contrasted, for example, with history, philosophy, or human affairs.

Nevertheless, *de facto* science articulates, in a self-conscious and methodologically explicit manner, the demands of objectivity over a staggering range of issues of natural fact, subjecting these issues continuously to the joint tests of theoretical coherence and observational fidelity. "It takes its starting points outside the mind in nature," writes C. C. Gillispie, "and winnows observations of events which it gathers under concepts, to be expressed mathematically if possible and tested experimentally by their success in predicting new events and suggesting new concepts."[3] This it does in a logically deliberate and progressively more general manner, thereby providing us with a comprehensive model of the ideal of objectivity itself, stretching our earlier conceptions of its potentialities, and pointing the way to new and as yet undreamed of embodiments in a variety of realms.

What I am saying may be put summarily as follows: Current science is continuous with other areas of life, and shares with them

[2] S. I. Benn and R. S. Peters, *Social Principles and the Democratic State* (London: George Allen and Unwin, Ltd., 1959), p. 22.
[3] C. C. Gillispie, *The Edge of Objectivity* (Princeton: Princeton University Press, 1960), p. 10.

the distinctive features of the rational quest. However, in institutionalizing this quest so as to subject an ever wider domain of claims to refined and systematic test, science has given us a new appreciation of reason itself. Since reason is, moreover, a moral as well as an intellectual notion, we have thereby been given also a new and enlarged vision of the moral standpoint—of responsibility in belief, embodied not only in a firm commitment to impartial principles by which one's own assertions are to be measured, but in a further commitment to making those principles ever more comprehensive and rigorous. Thus, though science has certainly provided us with new and critically important knowledge of man's surroundings and capacities, such enlightenment far from exhausts its human significance. A major aspect of such significance has been the moral import of science: its dynamic articulation of the impulse to responsible belief, and its suggestion of the hope of an increased rationality and responsibility in all realms of conduct and thought.

Such moral import has surely been too little appreciated. Yet it is, I think, true to say that the main twentieth-century scientific philosophies, in their rejection of post-Kantian speculative idealism, all celebrated this moral aspect of science in their varying idioms. Realism, to begin with, counterposed the simple stubbornness of perceived particulars to the attempt to spin a metaphysical yarn that would tell the tale of the universe, the latter attempt being fundamentally impatient with humble truths and hence irresponsible with regard to the Truth. The idealistic metaphysic, which construed the world as essentially mental, seemed to realists a piece of egotism founded on wishful thinking, and made possible only by the idealist's high-handedness with detailed realities independent of the will.

Pragmatism emphasized not the brute hardness of things transparently evident to consciousness, but rather the control of an organism's conceptions by its actions and their connected consequences in experience. It demanded that speculative abstractions be rejected as meaningless unless they could be reconstrued as predicting differential sensible outcomes of specified operations. It insisted, further, that truths acquired their warrant through publicly verifiable anticipations of the course of experience, contingent upon human transformations of the environment. To propound as true a belief protected from the hard test of experience flowing out of action is, from the pragmatist's standpoint, willful self-assertion or self-deception, and only secondarily irresponsible in dissipating the power to remedy avoidable evils and to render more lovely the scene of human life.

Logical positivism, finally, stressed the special place of language, logic, and system in mediating the control of our beliefs by observed phenomena. For the logical positivist, statements are cognitively meaningful if, and only if, they are verifiable—if, and only if, that is to say, it is clear from the language of these statements how observations might conceivably make a difference to their warrant. The observational import of a statement need not, of course, be borne on its face, but may accrue to it indirectly, through appropriate logical links with other relevant statements. However, a statement which lacks even indirect observational bearing can, for the positivist, have no cognitive meaning at all. To assert such a statement is irresponsible; it is to put forward for acceptance what one could never, even conceivably, have any experiential reason for holding true.

The cultural storm raised by positivism stemmed from its purported negativism, from its relegation of metaphysics, poetry, religion, and ethics to the limbo of the cognitively meaningless. What went largely unnoticed in the general indignation was the underlying moral impulse of positivism, the conviction that our assertions impose upon us the responsibility to satisfy relevant independent controls. The *unity of science* doctrine, urged by positivists, had, I should suggest, a similar moral motivation: to affirm the responsibilities of assertion no matter what the subject matter, to grant no holidays from such responsibilities for the humanities, politics, or the social sciences in particular, despite their strong capacities for arousing emotion and stimulating partisanship.

These twentieth-century scientific philosophies all had their special problems, to be sure. Realism, for example, had particular difficulty in accounting for errors and illusions—that is to say, for all those cases in which the stubbornness of particulars is not a sign of truth but rather a mask of falsehood. It seemed able to accommodate error only by admitting mediate ideational processes to separate realities from illusions, thus diluting its forthright appeal to the brute being of things as grasped in awareness. Pragmatism, in stressing from the beginning such mediate ideational processes construed in terms of the organism's biography of action and experience, ran the opposite risk—that of losing altogether the hard reality of independent things, and the simple vision of simple truths concerning them. Finally, logical positivism found the formulation of precise criteria of meaning to be a tantalizing business: set the criteria too high and you exclude perfectly respectable areas of natural science; set them too low and you include pseudo-science and superstition. The ironic course of positivism was, in fact, one of progressive liberalization to the point where virtually nothing could be denied cognitive meaning by reference to its criteria of observational control; its acid negativism had turned to water.

But if these philosophies all faced grave problems in the development of their several views of objective control over assertion, they never wavered on the fact of such control, nor on its central significance for the understanding of scientific knowledge and procedure. And they found post-Kantian idealism wanting precisely because it failed to give adequate acknowledgment to such control, and, consequently, failed to provide an adequate interpretation of the actuality of science. The general points were well put by C. I. Lewis:

> If the mind were the *only* condition of the thing as known, then the nature of the mind being specified, objects in general would be completely determined. . . .
>
> Idealism has often boggled over the fact that it could not deduce the particular content of experience and knowledge. The questions, "Why do I have just this experience? Why do I find just this reality and no other?" must have an answer. . . . The post-Kantian idealists . . . have either neglected this problem or, like Fichte, have said that it is no part of the business of philosophy to deduce the particular. But he fails to face the question: Granted the idealistic thesis, *can* the particular be deduced?* . . . . If the mathematician should tell us, "All the facts of physics can be deduced from the system of quaternions," but should reply to our request for a deduction of the law of gravitation by saying, "Particular physical facts are of no interest to the mathematician and no part of his business," we should draw our own conclusions. . . .

Unless the content of knowledge is recognized to have a condition independent of the mind, the peculiar significance of knowledge is likely to be lost. For the purpose of knowledge is to be true to something which is beyond it. Its intent is to be governed and dictated to in certain respects. It is a real act with a real purpose because it seeks something which it knows it may miss. If knowledge had no condition independent of the knowing act, would this be so?[4]

*Schelling, however, acknowledges the justice of the challenge and seeks to meet it—with amazing results. Starting from the Fichtean premise, $A=A$, he deduces eventually the electrical and magnetic properties of matter! *System d. transcendentalen Idealismus,* sämmt. Werke (1858), Bd. I, 3, pp. 444—450.

The insistence on independent and controlling conditions which define standards of responsibility for the "knowing act" unifies, indeed, the main scientifically oriented philosophies of the present century. Despite their individual variations, these philosophies have been inspired by the moral example of science in the realm of belief; in their several ways they have exalted the ideal of responsible control over assertion. In so doing, they have fed into and strengthened a common philosophy of science with independent roots as well, a philosophy which has attained the status of a standard view, largely shared by reflective scientists, technical philosophers, and the educated public alike, and laying great emphasis upon the objective features of scientific thought. It is to the prospects of this standard view that I wish mainly to address myself, for I believe that it is coming increasingly under fundamental attack.

The philosophical scene is, in this respect, undergoing a radical change indeed. For the standard view has not only been widely entrenched and long taken for granted; it has also, as we have seen, enjoyed the staunch support of the dominant scientifically oriented philosophies of our day. The current attacks thus challenge not only a firm set of habitual attitudes, but also the very opposition between science and speculative idealism, from which the scientifically minded philosophies have sprung. The attacks threaten further the

[4]Clarence Irving Lewis, *Mind and the World Order,* republication of first edition with corrections by the author (New York: Dover Publications, Inc., 1956), pp. 189—192. First published, New York: Charles Scribner's Sons, 1929. By permission of Andrew K. Lewis.

underlying moral motivation of these philosophies, their upholding of the ideal of responsibility in the sphere of belief as against willfulness, authoritarianism, and inertia. The issues are fundamental, indeed more fundamental than is generally realized, precisely because a powerful moral vision has implicitly been called into question. Nor is there any reasonable alternative to a critical confrontation of these issues, in the knowledge that enormously much is at stake. We must now seriously ask ourselves whether scientific objectivity is not, after all, an illusion, whether we have not, after all, been fundamentally mistaken in supposing empirical conceptions capable of responsible control by logic and experience. The question before us becomes, in short: How, if at all, is scientific objectivity possible?

In approaching this question, we shall begin by elaborating what has above been described as the "standard view" of science. Fundamentally, as we have seen, this view affirms the objectivity of science; more specifically, it understands science to be a systematic public enterprise, controlled by logic and by empirical fact, whose purpose it is to formulate the truth about the natural world. The truth primarily sought is general, expressed in laws of nature, which tell us what is always and everywhere the case. Observation, however, supplies the particular empirical facts, the hard phenomenal data which our lawlike hypotheses strive to encompass, and for which it is the ultimate purpose of such hypotheses to account.

Laws or general hypotheses may be ordered in a hierarchy of increasing generality of scope, but a basic distinction is, in any event, to be drawn between observational or experimental laws on the one hand, and theoretical laws on the other. Generalizing upon the data accessible to the senses, observational laws are couched in the language of observation and make reference to perceived things and processes. Theoretical laws, by contrast, are expressed in a more abstract idiom and typically postulate unobservable elements and functions; unlike observational laws, they cannot be subjected to the test of direct inspection or experiment. Their function is not to generalize observed phenomena, but rather to explain the laws which themselves generalize the phenomena. This they do by yielding such laws as deductive consequences of their own abstract postulations. They are, of course, indirectly testable by observation, for should one of their lawlike consequences break down on the level of experiment, such failure would count against them. However, they serve primarily to help relate diverse observational laws suitably within a comprehensive deductive scheme, and they are evaluated not only by their empirical yield but also by their simplicity, their intellectual familiarity, their accessibility to preferred models, and their manageability. They are, to be sure, also applied in the explanation of particular occurrences and in the solution of problems of prediction and control.

Any two theories of the same domain of phenomena may be compared to see if either is superior in accounting for the relevant empirical facts or, if equivalent on this score, if either surpasses the other in simplicity or convenience, etc. A hypothesis that does not itself clash with experience may yet be given up in favor of an alternative hypothesis that explains more facts or is simpler, or easier to handle. A given law may be absorbed into another, more general law by a process of reduction, through which it is shown to follow deductively from the more general law under plausible auxiliary assumptions.

When one hypothesis is superseded by another, the genuine facts it had purported to account for are not inevitably lost; they are typically passed on to its successor, which conserves them as it reaches out to embrace additional facts. Thus it is that science can be cumulative at the observational or experimental level, despite its lack of cumulativeness at the theoretical level; it strives always, and through varying theories, to save the phenomena while adding to them. And in the case of reduction, a reduced law is itself conserved, *in toto,* as a special consequence of its more general successor. Throughout the apparent flux of changing scientific beliefs, then, there is a solid growth of knowledge which represents progress in empirical understanding. Underlying historical changes of theory, there is, moreover, a constancy of logic and method, which unifies each scientific age with that which preceded it and with that which is yet to follow.

Such constancy comprises not merely the canons of formal deduction, but also those criteria by which hypotheses are confronted with the test of experience and subjected to comparative evaluation. We do not, surely, have explicit and general formulations of such criteria at the present time. But they are embodied clearly enough in scientific practice to enable communication and agreement in a wide variety of specific cases. Such communication and consensus indicate that there is a codifiable methodology underlying the conduct of the scientific enterprise. It is a methodology by which beliefs are objectively evaluated and exchanged, rather than an organon of discovery or theoretical invention. Yet it is this methodology which makes possible the cumulative growth of tested scientific knowledge as a public possession.

Now the public character of scientific procedure is not simply a matter of the free interchange of ideas. It is intimately related to the critical testing of beliefs, in the following way: If I put forward a hypothesis in scientific spirit, I suppose from the outset that I may be wrong, by independent tests to which I am prepared to submit my proposal. I suppose, in other words, that my present hypothesis is not to be prejudged as correct during the process of testing; I thus acknowledge that disagreement with respect to my proposal is no bar to further communication, nor indeed to agreement on the test itself. Indeed, from the latter sort of communication and agreement, consensus on my proposal may eventually grow. Further, insofar as testing involves an appeal to facts disclosed in common observation of things, I suppose that the same things can be observed from different perspectives, and consensus on observation reached without presupposing agreement on relevant theory. In sum, I acknowledge the possibility of common discourse with those who may differ with me in opinion, and assume shared access to an observed world with others who may be differently located or otherwise constituted than I. The methodological publicity of science involves the assumption that differing persons may yet talk intelligibly to one another, that they may observe together the phenomena bearing critically on issues which divide them, and that they may thus join in the testing of disputed conceptions in an effort to seek resolution.

And, indeed, it seems undeniable that resolution often occurs. In the free community of scientific discourse, untrammeled by doctrinal bounds, convergence of opinion yet takes place. It seems, in fact, often to be conserved and progressively expanded, at least on the experimental level, if not on the higher level of theoretical ideas. With no attempt to shape opinion to advance specifications, with free access to evidence and no prior limitation on the community of discourse, opinion nevertheless forms and crystallizes. Does this not provide reasonable grounds for assuming that reality itself, that is to say, a world independent of human wish and will, progressively constrains our scientific beliefs? An interpretation of this sort has been colorfully expressed by Charles Peirce:

> Different minds may set out with the most antagonistic views, but the progress of investigation carries them by a force outside of themselves to one and the same conclusion. This activity of thought by which we are carried, not where we wish, but to a foreordained goal, is like the operation of destiny. No modification of the point of view taken, no selection of other facts for study, no natural bent of mind even, can enable a man to escape the predestinate opinion. This great law is embodied in the conception of truth and reality. The opinion which is fated to be ultimately agreed to by all who investigate is what we mean by the truth, and the object represented in this opinion is the real. That is the way I would explain reality.[5]

The reality thus revealed under the methodological publicity of scientific method is, moreover, a reality in which we are ourselves but limited natural elements. Our wishes and perceptions have not made this reality, but have sprung up within it as functions of organic development in a small corner of the universe of nature. Objectivity is not only, as we have seen, a fundamental feature of scientific method; the ontological vision in which it culminates is the vision of a universe of objects with independent existences and careers, within which scientific inquiry represents but one region of connected happening and striving. In short, for the standard view I

[5] Charles S. Peirce, "How to Make Our Ideas Clear," *Popular Science Monthly*, XII (1878), 286–302. Reprinted in Charles S. Peirce, *Essays in the Philosophy of Science*, ed. Vincent Tomas, "The American Heritage Series" no. 17 (Indianapolis: The Bobbs-Merrill Co., Inc., 1957) pp. 31–56; for the passage cited, see Tomas, pp. 53–54.

have been describing, objectivity is the end, as well as the beginning, of wisdom.

Recent attacks against this standard view have been launched from various directions. They have varied also in scope and precision, and their larger strategic import has not always been evident, even to the combatants themselves. Yet, taken together, these attacks add up, in my opinion, to a massive threat to the very possibility of objective science. Uncoordinated as they are, they have already subtly altered the balance of philosophical forces, exposing to danger the strongest positions of the objectivist viewpoint.

In dealing with this general situation, I shall divide the field into three main sectors, embracing, respectively, issues of observation, of meaning, and of scientific change. In the three lectures immediately following, I shall address myself to each of these sectors in turn, in an effort to evaluate the status of the objective ideal within it, in the light of the attacks to which it has there been exposed. My purpose is, in a broad sense, logical rather than historical; I shall therefore feel free to abstract from the detail of historical developments, in order to emphasize considerations which I take to have fundamental logical bearing on the fortunes of objectivity. In the remainder of the present lecture, I shall sketch some of these considerations, and try to provide an initial sense of the predicament of the standard view.

Consider then, first, the idea that observation supplies us with hard data independent of our conceptions and assertions, data by which, indeed, our conceptions and assertions are controlled. C. I. Lewis has expressed the point in terms of "the given," but the underlying notion is quite widespread and is embedded in the standard view. "The given," he writes, ". . . is what remains unaltered, no matter what our interests, no matter how we think or conceive."[6] Conception, thought, and interest may produce varying interpretations of the given, but they cannot create or change it. Strip away all interpretation contributed by the mind and you will find underneath a somewhat which is what it is as presented to sense, and which must be accepted as such, though estimates of its import may differ. In-

[6] *Mind and the World Order*, p. 52.

terpretation must, in short, be interpretation of something, and that something must itself be independent of interpretation if the interpretive process is not to collapse into arbitrariness.

Now it is clear that such a view promises the advantage of providing an external standard for the testing and evaluation of our thought, but is it a tenable view, and can it fulfill the promise? Can we, to begin with, accept the supposition that an unalterable observable somewhat underlies all conceptualization, interpretation, and valuation in experience? Can we even begin to imagine what it would be like to de-categorize our thought and strip away all interpretation, so as to enable the critical confrontation with a pure given, if there be such a thing? There seems to be good psychological reason to suppose that observation is not at all a bare apprehension of pure sense content, but rather an active process in which we anticipate, interpret, and structure in advance what is to be seen. There are, indeed, things right in front of our eyes that we fail to see, and things we see of which we have only the faintest clues in context—guided by expectation, we even see what is not there at all, as any proofreader knows. But if observation is never conceptually neutral, if it cannot occur without expectation and schematization, then stripping away all interpretation leaves nothing at all. And if nothing at all remains, what is there to provide an independent check on such interpretation?

A related line of thought may be developed, which presupposes no special psychological notions concerning the nature of observation, but yields what appears to be a paradox from the vantage point of the standard view. We may call it the *paradox of categorization* and explain it as follows: If my categories of thought determine what I observe, then what I observe provides no independent control over my thought. On the other hand, if my categories of thought do not determine what I observe, then what I observe must be uncategorized, that is to say, formless and nondescript—hence again incapable of providing any test of my thought. So in neither case is it possible for observation, be it what it may, to provide any independent control over thought. Nor does it help to suggest that my thought categories determine only in part, or in certain respects, what I observe. For any part or aspect thus determined can obviously provide no independent control over my thought, while every part or aspect not thus determined must remain for me formless and ineffable, hence, in particular, incapable of categorization in its bearing on my thought. Again the possibility of control over thought by observation seems to have vanished. Observation contaminated by thought yields circular tests; observation uncontaminated by thought yields no tests at all.

Now consider that, on the standard view, people with different theoretical beliefs may observe the same things; shared access to a common world is taken for granted. Yet, unless this common world is to be construed as pure formless given, hence too fluid to yield shareable objects, it must be conceived, on the contrary, as structured by particular categories of thought. It seems to follow, then, that a difference in categories destroys the common character of the world—implying, in fact, a difference in things observed. A small child, for example, sees a hard, table-shaped object resistant to his push, and capable of supporting small items placed on it, whereas a physicist sees a peculiar swarm of electrons obeying complex physical laws. More importantly, scientists with different theories categorize the objects of their observation in correspondingly different ways, and must therefore, in a critical sense, be said to see different things.

N. R. Hanson has suggested a view of this sort in certain passages of his *Patterns of Discovery,* in which he stresses the dependence of seeing upon theory, and argues against the notion that theoretical differences in a given domain must be attributed simply to differing interpretations of the same observational data. "There is a sense, then," he writes, "in which seeing is a 'theory-laden' undertaking. Observation of x is shaped by prior knowledge of x."[7] The visitor to the physicist's laboratory "must learn some physics before he can see what the physicist sees. . . . The infant and the layman can see: they are not blind. But they cannot see what the physicist sees; they are blind to what he sees."[8] To suppose "that Kepler and Tycho see the same thing at dawn just because their eyes are similarly

---

[7] Norwood Russell Hanson, *Patterns of Discovery* (Cambridge at the University Press, 1958), p. 19.

[8] *Ibid.,* p. 17.

affected is an elementary mistake. There is a difference between a physical state and a visual experience."[9] Controversy in scientific research is too deep-seated, argues Hanson, to be explained simply by appeal to differing interpretations of the same data; divisions at the theoretical level cut down through the level of data as well. "It is the sense in which Tycho and Kepler do not observe the same thing," he writes, "which must be grasped if one is to understand disagreements within microphysics."[10]

Such a line of thought seems again, however, from the perspective of the standard view, to lead to paradox. For if seeing is indeed theory-laden in the sense described, then proponents of two different theories cannot observe the same things in an effort to resolve their differences; they share no neutral observations capable of deciding between them. To judge one theory as superior to the other by appeal to observation is always doomed, therefore, to beg the very question at issue. We may call this the *paradox of common observation*. It has the effect of isolating each scientist within an observed world consonant with his theoretical beliefs.

It cannot be denied, of course, that scientists who differ theoretically may yet share a common observational or experimental vocabulary. This is indeed the basis for the differentiation made, in the standard view, between observational and theoretical levels of scientific discourse. It might thus be argued that even if theoretical differences prevent observation of common things, yet such differences allow for a shared discourse, based on the communication of common meanings. But can this be so? To adopt a new theory is, after all, to employ it not only in rethinking the phenomena, but also in reassigning the roles of relevant descriptive terms and in recasting familiar definitions and explanations. Even where there is no explicit revision of the latter sorts, a newly adopted theory alters the background of assumptions by reference to which every relevant term must be located. A new theory thus, in effect, provides new senses for old observational terms by incorporating them within a new framework of assumptions and meanings.

Thomas S. Kuhn, discussing the possible derivation of Newtonian

[9]*Ibid.*, p. 8   [10]*Ibid.*, p. 18.

from relativistic dynamics, argues that such derivation is, strictly speaking, impossible, for the meanings of critical parameters such as space, time, and mass undergo change under alteration of theory, so that the ostensibly Newtonian formulas derived from relativity theory are actually Einsteinian correlates of such formulas, with differing meaning. "The physical referents of these Einsteinian concepts," writes Kuhn,

> are by no means identical with those of the Newtonian concepts that bear the same name. (Newtonian mass is conserved; Einsteinian is convertible with energy. Only at low relative velocities may the two be measured in the same way, and even then they must not be conceived to be the same.) Unless we change the definitions of the variables . . . , the statements we have derived are not Newtonian. If we do change them, we cannot properly be said to have *derived* Newton's Laws, at least not in any sense of "derive" now generally recognized. . . . [The argument] has not, that is, shown Newton's Laws to be a limiting case of Einstein's. For in the passage to the limit it is not only the forms of the laws that have changed. Simultaneously we have had to alter the fundamental structural elements of which the universe to which they apply is composed.
>
> . . . Just because it did not involve the introduction of additional objects or concepts, the transition from Newtonian to Einsteinian mechanics illustrates with particular clarity the scientific revolution as a displacement of the conceptual network through which scientists view the world.[11]

Such conceptual displacement, if it is conceived as affecting observational as well as theoretical notions, means that the ostensible sharing of observational terms by theoretical opponents is really a delusion. There are perhaps common sounds but no common meanings. There can thus be no intelligible converse between scientists of differing theoretical persuasions. To understand another's apparently observational or experimental references, we must first enter into his theoretical thought-world.

It seems to follow, further, that we cannot literally speak of alter-

[11] Reprinted from *The Structure of Scientific Revolutions* by Thomas S. Kuhn by permission of The University of Chicago Press (Chicago and London: The University of Chicago Press, 1962), p. 101. Copyright under the International Copyright Union by The University of Chicago. All rights reserved.

native theories *of the same domain,* nor of comparing these theories to see which gives a better account of the empirical facts within this domain. For there is not, and there cannot be, a neutral account of the domain in question, since the observational derivations of each theory differ in meaning from those of the other, no matter how similar they are simply as sound patterns. Nor can one law really be absorbed into another through a process of reduction, nor observational content passed on from one theory to its successor, for crucial meaning changes have occurred in the process of transfer. We have here another paradox, the *paradox of common language,* and its upshot is that there can be no real community of science in any sense approximating that of the standard view, no comparison of theories with respect to their observational content, no reduction of one theory to another, and no cumulative growth of knowledge, at least in the standard sense. The scientist is now effectively isolated within his own system of meanings as well as within his own universe of observed things.

The breakdown of observational community and of the community of meaning, and the consequent rejection of cumulativeness seem to remove all sense from the notion of a rational progression of scientific viewpoints from age to age. If contemporary theoretical alternatives cannot be compared and evaluated with respect to their factual accuracy and comprehensiveness, neither can succeeding theoretical alternatives be thus compared and evaluated. The genesis of a new theory cannot be backed up by an established methodology of justification, such as is presupposed by the standard view; indeed, the very distinction between justification and discovery breaks down as well.

Supplementary arguments of a historical sort are, furthermore, available, to the effect that appeals to evidence are not generally decisive in the process of theoretical transition. Kuhn thus argues, on the basis of historical examples, that before the proponents of differing scientific paradigms "can hope to communicate fully, one group or the other must experience the conversion that we have been calling a paradigm shift. Just because it is a transition between incommensurables, the transition between competing paradigms cannot be made a step at a time, forced by logic and neutral expe-

rience."[12] He further writes, "No process yet disclosed by the historical study of scientific development at all resembles the methodological stereotype of falsification by direct comparison with nature."[13] And Michael Polanyi, emphasizing the "intuition of rationality in nature,"[14] argues, on the basis of his interpretation of scientific history, that knowledge in science is personal, committing us
"passionately and far beyond our comprehension, to a vision of reality. Of this responsibility we cannot divest ourselves by setting up
objective criteria of verifiability—or falsifiability, or testability, or
what you will."[15] The general conclusion to which we appear to be
driven is that adoption of a new scientific theory is an intuitive or
mystical affair, a matter for psychological description primarily,
rather than for logical and methodological codification.

The categories of logic and methodology indeed do give way to
those of psychology, and even of politics and religion, in certain recent historical accounts of science. Kuhn thus speaks of the "gestalt
switch"[16] and Polanyi of "passionate, personal, human appraisals
of theories."[17] Kuhn has also employed a political vocabulary in his
descriptions: he speaks of crises and revolutions, and of the victory
which results in a rewriting of history so as to make progress seem
inevitable. And we have already noted the reference to conversion;
we now learn that converts to a new theory are made through persuasion, that the light is eventually seen, or, if not, that the new
theory gains ascendancy when the opposing generation dies in the
wilderness. In Planck's words, cited by Kuhn, "A new scientific
truth does not triumph by convincing its opponents and making

---

[12] *The Structure of Scientific Revolutions,* p. 149.

[13] *Ibid.,* p. 77.

[14] Michael Polanyi, *Personal Knowledge* (Chicago: The University of Chicago Press, and London: Routledge & Kegan Paul, Ltd., 1958; revised
edition, 1962), p. 16. Reprinted by permission of the publisher. Reissued,
Harper Torchbook edition (New York and Evanston: Harper & Bros.,
1964).

[15] *Ibid.,* p. 64.

[16] *The Structure of Scientific Revolutions,* p. 149.

[17] *Personal Knowledge,* p. 15.

them see the light, but rather because its opponents eventually die, and a new generation grows up that is familiar with it."[18]

Finally, with cumulativeness gone, the concept of convergence of belief fails, and with it the Peircean notion of reality as progressively revealed through scientific advance. For there is no scientific advance by standard criteria, only the rivalry of theoretical viewpoints and the replacement of some by others. Reality is gone as an independent factor; each viewpoint creates its own reality. Paradigms, for Kuhn, are not only "constitutive of science"; there is a sense, he argues, "in which they are constitutive of nature as well."[19]

But now see how far we have come from the standard view. Independent and public controls are no more, communication has failed, the common universe of things is a delusion, reality itself is made by the scientist rather than discovered by him. In place of a community of rational men following objective procedures in the pursuit of truth, we have a set of isolated monads, within each of which belief forms without systematic constraints.

I cannot, myself, believe that this bleak picture, representing an extravagant idealism, is true. In fact, it seems to me a *reductio ad absurdum* of the reasonings from which it flows. But it is easier, of course, to say this than to pinpoint the places at which these reasonings go astray. The latter task is what I shall undertake in the following three lectures. In the next, I examine the attacks on the objectivity of observation.

[18] Max Planck, *Scientific Autobiography and Other Papers*, tr. Frank Gaynor (New York: Philosophical Library, 1949), pp. 33–34, cited in Kuhn, p. 150.

[19] *The Structure of Scientific Revolutions*, p. 109.

2
Observation and Objectivity

As REMARKED at the close of the last lecture, the reasoned rejection of objectivity seems to involve a *reductio ad absurdum,* in fact, a form of self-refutation. For objectivity is relevant to all statement which purports to make a claim, to rest on argument, to appeal to evidence. Science, as I have urged, is not uniquely subject to the demands of objectivity; rather, it institutionalizes such demands in the most systematic and explicit manner. But to put forth *any* claim with seriousness is to presuppose commitment to the view that evaluation is possible, and that it favors acceptance; it is to indicate one's readiness to support the claim in fair argument, as being correct or true or proper. For this reason, the particular claim that evaluation is a myth and fair argument a delusion is obviously self-destructive. If it is true, there can be no reason to accept it; in fact, if it is true, its own truth is unintelligible: what can truth mean when no evaluative standard is allowed to separate it from falsehood? And indeed there is a striking self-contradictoriness in the effort to persuade others by argument that communication, and hence argument, is impossible; in appeal to the facts about observation in order to deny that commonly observable facts exist; in arguing from the hard realities of the history of science to the conclusion that reality is not discovered but made by the scientist. To accept these claims is to deny all force to the arguments brought forward for them. Acceptance implies that we are free to assert what we will and hence, in particular, to reject them.

Nevertheless, we cannot let the matter rest here. For the claims in question are supported by detailed considerations to which we ourselves are inclined to assent, at least initially. They appeal to purported facts of psychology, language, and history which have a broader base than any partisan philosophical construction erected upon them, and they proceed by a dialectic which is generally compelling. The self-refutation in which they culminate is thus indicative of serious trouble in our own house. Unless we find some way of blocking the paradoxical conclusions based upon them, we are ourselves committed to these very conclusions. Nor will a mindless objectivism, loudly proclaimed, remove the difficulty. What is wanted is a serious examination of the causes of such difficulty, an analysis of the multiple considerations upon which it feeds. If such an analysis eventually leads us back to objectivism, it will be an objectivism that can be sustained because its conceptual context has been suitably informed and thereby transformed.

I turn, then, to the task of such analysis, and address myself first to the problem of objectivity in observation. In its broadest terms, the fundamental issue may be stated as follows: Observation needs to be construed as independent of conceptualization if conceptualization is not to be simply arbitrary; yet, it cannot plausibly be thought to be independent of conceptualization. On the contrary, it is shot through with interpretation, expectation, and wish. Were it not so, indeed, we should be powerless to take hold of anything in experience: equally receptive to everything in awareness and uniformly undiscriminating, we could not properly be said to observe anything at all; at best we should confront a flat and undifferentiated given incapable of providing any control over our thought. So, on the one hand, observation must be independent, and, on the opposite hand, it cannot be. To suppose it is independent commits us to an implausibly pure observational given, and makes a mystery of observational control over thought. To suppose, on the other hand, that it is not independent commits us rather to the view that apparent observational control is always circular and hence incapable of restricting the arbitrariness of conception; it further commits us to the impossibility of common observation across the barrier of conceptual differences. In any event, the standard notion of observation as providing objective control over conception needs to be abandoned.

It will be instructive to look more closely at this problem in the context of a philosophy which strongly defends the given as an independent control over conceptualization—I have in mind the philosophy of C. I. Lewis, as set forth in his important book of 1929, *Mind and the World Order*.[1] As we have earlier noted, Lewis insists that "unless the content of knowledge is recognized to have a condition independent of the mind, the peculiar significance of knowledge is likely to be lost" (p. 191). I shall interpret his doctrine of the given in the light of this insistence, and shall elaborate my own treatment of observational objectivity through a dialectical interchange with Lewis' views.

For Lewis, "The two elements to be distinguished in knowledge are the concept, which is the product of the activity of thought, and the sensuously given, which is independent of such activity" (p. 37). It is a principal thesis of Lewis concerning their relationship that "The pure concept and the content of the given are mutually independent; neither limits the other" (p. 37). Conceptual interpretation of the given, however, gives rise to empirical truth or objective knowledge. Such knowledge is always predictive and, at best, probable, for the concept denotes an object and an object is "never a momentarily given as such, but is some temporally-extended pattern of actual and possible experience" (p. 37). To apply a concept to what is momentarily given is thus to interpret this given as belonging to a larger pattern of experience, at least part of which extends into the future. The further course of experience thus provides a test of my present assignment of a concept to what is now given to me.

In discussing his distinction between the concept and the given,

[1] All following page references to Lewis in the text refer to the Dover edition of 1956 (see note 4 above, p. 7.)

Lewis maintains that some such contrast, between "the immediate data, such as those of sense, which are presented or given to the mind, and a form, construction, or interpretation, which represents the activity of thought. . . . is one of the oldest and most universal of philosophic insights" (p. 38). Its central role in Lewis' thought is indicated by his assertion that:

> To suppress it altogether, would be to betray obvious and funda- mental characteristics of experience. If there be no datum given to the mind, then knowledge must be contentless and arbitrary; there would be nothing which it must be true to. And if there be no inter- pretation or construction which the mind itself imposes, then thought is rendered superfluous, the possibility of error becomes in- explicable, and the distinction of true and false is in danger of be- coming meaningless [pp. 38–39].

In order to provide thought with content and so to restrict its arbitrariness, Lewis here requires data given to the mind—data, therefore, which thought cannot form but to which it must itself be true. Does this not imply, however, that the mind is capable of ap- prehending the data of sense as given to it, prior to its own inter- pretation and construction? And does not such a supposition con- flict with considerations that weigh overwhelmingly against a pure observational receptivity to sense content? Does it not contradict the active, constructive, and anticipatory involvement of the ob- server in all observation?

Regarding such involvement, we may well consider, for example, the words of E. H. Gombrich:

> Without some starting point, some initial schema, we could never get hold of the flux of experience. Without categories, we could not sort our impressions. Paradoxically, it has turned out that it matters relatively little what these first categories are. We can always ad- just them according to need. Indeed, if the schema remains loose and flexible, such initial vagueness may prove not a hindrance but a help. An entirely fluid system would no longer serve its purpose; it could not register facts because it would lack pigeonholes. But how we arrange the first filing system is not very relevant.[2]

---

[2] *Art and Illusion* by E. H. Gombrich. The A. W. Mellon Lectures in the Fine Arts, 1956. 2nd edition, copyright 1960. Bollingen Series XXX.5. Pantheon Books. By permission of the Trustees of the National Gallery of Art, Washington, and the Bollingen Foundation, p. 88.

Accordingly, Gombrich stresses the related notions "schema" and "correction" in all perception, the implication being that we can never separate data from construction, never observe in the sense of simply registering unconceptualized content, to be interpreted in a later phase. It follows, in sum, that observation cannot be independent of conception, if independence requires such simple registration of unconceptualized content.

To this line of argument, however, Lewis has an answer. For there is a difference between the notion of an actual separation of observational and conceptual processes, on the one hand, and the notion of a theoretical separation of observational and conceptual components in experience, on the other. To formulate the problem, as we have hitherto done, simply in terms of the independence of observation from conceptualization, masks this critical difference. It is similarly ambiguous to ask whether all interpretation contributed by the mind can be stripped away, leaving a pure sensuous given beneath. For, in one sense of this question, it asks whether thought can actually be decategorized in such a way as to comprise a purely passive apprehension of sensory content. In another sense of the question, it asks, rather, whether thought can be theoretically decomposed into the two components of given and interpretation, so that analytic attention may be concentrated upon the former in abstraction from the latter. It is surely consistent to hold that, in actual thought, given and interpretation are always indissolubly linked, and yet to believe that they may be analytically distinguished for theoretical purposes. Such, indeed, is Lewis' view. In speaking of the given, he is not committed to conceiving the mind as first passively accepting the given and then, in a second stage, processing it. The mind may be as thoroughly active as you like; it may reach out to grasp the data—perhaps more properly called "taken" than "given"—and may begin its processing from the very first instant of contact with them; yet its own active contribution may remain analytically distinguishable from the data themselves.

Lewis is, in fact, concerned not with an actual separation of the processes of observation and conceptualization but rather with a theoretical analysis of different components of experience. The given, as he conceives it, "is certainly an abstraction," and (barring the dubious possibility of pure esthesis) "never exists in isolation in any experience or state of consciousness" (p. 54). It "is *in,* not before experience" (p. 55). Indeed, as he emphatically declares:

> A state of intuition utterly unqualified by thought is a figment of the metaphysical imagination, satisfactory only to those who are willing to substitute a dubious hypothesis for the analysis of knowledge as we find it. The given is admittedly an excised element or abstraction; all that is here claimed is that it is not an "unreal" abstraction, but an identifiable constituent in experience [p. 66].

Nor can abstraction, in general, be condemned, for "the condemnation of abstractions," as Lewis puts it, "is the condemnation of thought itself" (p. 55).

The *independence* of the given is not, then, to be understood as the literal *isolation* of the given: it should not be taken as implying that there are states of pure observational receptivity which altogether precede the processes of interpretation. Nor does it entail the practical possibility of removing altogether the encrustation of concepts, so as to discover the core of the given, in a moment of sheer apprehension. Lewis' construal of the independence of the given thus survives the counterargument based on the active nature of the mind and, in particular, of the mind's observation. He still needs to tell us, however, how to locate the given analytically for theoretical purposes, and how positively to interpret its independence. If we cannot expect to sunder it from interpretation in actuality, we must at least be able to distinguish and identify it abstractly, as an embedded constituent of experience.

This we can do, according to Lewis, by looking first to the specifically sensuous aspect of experience, and then, more importantly, to the character of unalterability by thought, the given being "unaffected by any change of mental attitude or interest" (p. 66). Lewis provides the following illustration:

> At the moment, I have a fountain pen in my hand. When I so describe this item of my present experience, I make use of terms whose meaning I have learned. Correlatively I abstract this item from the total field of my present consciousness and relate it to what is not just now present in ways which I have learned and which reflect modes of action which I have acquired. . . . There is, to be

sure, something in the character of this thing as a merely presented colligation of sense-qualities which is for me the clue to this classification or meaning; but that just this complex of qualities should be due to a "pen" character of the object is something which has been acquired. Yet what I refer to as "the given" in this experience is, in broad terms, qualitatively no different than it would be if I were an infant or an ignorant savage.

Again, suppose my present interest to be slightly altered. I might then describe this object which is in my hand as "a cylinder" or "hard rubber" or "a poor buy." In each case the thing is somewhat differently related in my mind, and the connoted modes of my possible behavior toward it, and my further experience of it, are different. Something called "given" remains constant, but its character as sign, its classification, and its relation to other things and to action are differently taken [pp. 49–50].

Such constancy of the given through variation of interest is, moreover, concretely marked by certain limitations on my conceptualization. "I can apprehend this thing as pen or rubber or cylinder, but I cannot, by taking thought, discover it as paper or soft or cubical" (p. 52). The given is thus identifiable as "what remains unaltered, no matter what our interests, no matter how we think or conceive" (p. 52).

Thus identified by its unalterability, the given can, however, not be described *as such,* says Lewis,

because in describing it, in whatever fashion, we qualify it by bringing it under some category or other, select from it, emphasize aspects of it, and relate it in particular and avoidable ways. . . . So that in a sense the given is ineffable, always. It is that which remains untouched and unaltered, however it is construed by thought. Yet no one but a philosopher could for a moment deny this immediate presence in consciousness of that which no activity of thought can create or alter [pp. 52–53].

It is clear why Lewis wants the given to be sensuous and unalterable: to provide a phenomenal rock to which all our knowledge can be safely moored. The sensuous character of the given is accordingly not to be thought of as restricting it to sense data correlated with particular sense organs; the motivation is epistemological rather than psychological, and for epistemology a broader notion of the sensuous is required. Thus, for example, "the pleasantness or fearfulness of a thing may be as un-get-overable as its brightness or loudness" (p. 57). And, in general, "It is the brute-fact element in perception, illusion and dream (without antecedent distinction) which is intended" (p. 57).

The presented qualitative character of the given—what Lewis calls "the quale"—is, moreover, "directly intuited . . . and is not the subject of any possible error because it is purely subjective" (p. 121). Its subjectivity consists in the fact that it is wholly discoverable in the single experience, unlike the property of an object—the manifestation of a property always transcending the single experience and hence opening the door to the possibility of objective error in its ascription. Because the quality of the given is error-free, according to Lewis, it must also be ineffable, for description always introduces the potentiality of uncertainty and mistake. Unalterable, directly intuited, ineffable, and certain, the given provides, then, for Lewis, a fixed control over conceptualization, and thus a firm basis for the objectivity of knowledge. The motivation for this view is, I believe, understandable; but I believe also that it suffers from grave difficulties and cannot, in the end, be sustained. There are psychological considerations which contradict it, and serious internal troubles which it cannot overcome.

Looking first to the psychological considerations, it seems to me the evidence denies that anything plausibly construed as sensuously given is unalterable by change of mental attitude or interest. Indeed, the evidence points overwhelmingly in the opposite direction, to show that belief, expectation, and set, exercise an enormous role in determining the quality of the sensuous given. Now Lewis acknowledges, as we have seen, that the given is always *embedded* in conceptualization; he holds it nevertheless immune to alteration by conceptual change. It is this alleged immunity which, I believe, runs counter to the psychological facts.

Let me call again on Gombrich, who, in his book *Art and Illusion,* charmingly presents the gist of these facts as taken from the laboratory, from the history of art, and from everyday life. An anecdote from his personal experience recounted in this book provides a striking initial illustration. During the war, he worked in the Monitoring Service of the British Broadcasting Corporation, listening to "radio transmissions from friend and foe," and it was here that he realized, as he puts it, "the importance of guided projection in our understanding of symbolic material." His report is worth quoting in detail:

> Some of the transmissions which interested us most were often barely audible, and it became quite an art, or even a sport, to interpret the few whiffs of speech sound that were all we really had on the wax cylinders on which these broadcasts had been recorded. It was then we learned to what an extent our knowledge and expectations influence our hearing. You had to know what might be said in order to hear what was said. More exactly, you selected from your knowledge of possibilities certain word combinations and tried projecting them into the noises heard. The problem then was a twofold one—to think of possibilities and to retain one's critical faculty. Anyone whose imagination ran away with him, who could hear any words—as Leonardo could in the sound of bells—could not play that game. You had to keep your projection flexible, to remain willing to try out fresh alternatives, and to admit the possibility of defeat. For this was the most striking experience of all: once your expectation was firmly set and your conviction settled, you ceased to be aware of your own activity, the noises appeared to fall into place and to be transformed into the expected words. So strong was this effect of suggestion that we made it a practice never to tell a colleague our own interpretation if we wanted him to test it. Expectation created illusion.[3]

Comparable cases for seeing are available in abundance. McClelland and Atkinson, in an interesting study,[4] measured the effect of hunger upon the images projected by subjects onto a blank screen. After a preliminary phase in which faint images were actually shown on the screen, the experimenters exposed blanks but asked for description of the things shown; they found that food-related responses, and even the size of imagined food objects, increased with the increase of hunger. Commenting on another sort of case, Gombrich recalls the work of conjurers who arouse a train of expectations whereupon our imagination completes the picture. He

[3] E. H. Gombrich, *ibid.*, p. 204.

[4] D. C. McClelland and J. W. Atkinson, "The Projective Expression of Needs: I. The Effect of Different Intensities of the Hunger Drive on Perception," *Journal of Psychology* (1948), *25*, pp. 205–222.

reminds us further that "Intelligence officers intent on the reading of aerial reconnaissance photographs, X-ray specialists basing a
diagnosis on the faintest of shadows visible in a tissue, learn in a
hard school how often 'believing is seeing' and how important it
therefore is to keep their hypothesis flexible."[5]

Nor should these points be taken as applying only in specialized
contexts. On the contrary, our ordinary perceptions are normally
filtered through attitudes and filled out by expectation. Sherif and
Cantril review some of the experimental work on selective influences in perception, beginning with Külpe's study (1904) of the
effect of instructions on manner of performance. "In the Külpe line
of experiments," they write:

> the "set" is produced by the instructions of the experimenter. But
> the internal factors ("set" in this case) need not always be due to
> instructional set. In more natural settings some motivational stress,
> some social pressure, or some established norm in the individual
> may and does produce the "set" or "attitude" with which the stim-
> ulus field is observed. Take, for example, the case of the hungry
> man looking for bread or the case of a lover waiting in a crowd for
> his sweetheart.[6]

Expectation figures, too, in everyday processes of identification,
which proceed on the basis of cues normally sufficient to select the
objects in question; when these cues fail in fact, we tend anyway
to see them as having succeeded. Bruner, Goodnow, and Austin
discuss this point as follows:

> A bird has wings and bill and feathers and characteristic legs.
> But the whole ensemble of features is not necessary for making the
> correct identification of the creature as a bird. If it has wings and
> feathers, the bill and legs are highly predictable. In coding or cate-
> gorizing the environment, one builds up an expectancy of all of these
> features being present together. It is this unitary conception that has
> the configurational or Gestalt property of "birdness." Indeed, once
> a configuration has been established and the object is being identi-
> fied in terms of configurational attributes, the perceiver will tend

[5] *Art and Illusion,* pp. 210–211.

[6] Muzafer Sherif and Hadley Cantril, *The Psychology of Ego-Involvements*
(New York: John Wiley & Sons, Inc., 1947), p. 32.

to "rectify" or "normalize" any of the original defining attributes that deviate from expectancy. Missing attributes are "filled in" . . . , reversals righted . . . , colors assimilated to expectancy. . . .[7]

If we take schemata, with Scheerer, to include "the formation of frames of reference and mental sets, the development of both generalizations and concepts in their preverbal form, and the carrying over of implicit organizing principles from one problem-solving situation to another," the lesson of psychology and everyday experience would indeed seem to be that "such schemata can exercise a striking influence on individual contents of perception and thought."[8] Such considerations seem to me to refute any doctrine of the given, such as Lewis', which holds it to be "that which no activity of thought can create or alter" (p. 53). The given, it would seem, is not only *embedded in* conceptualization; it is *alterable by* conceptualization. If the independence of the given does not entail its isolation, neither does it, then, entail its fixity.

Lewis' view suffers also from various internal difficulties. Recall his saying, "I can apprehend this thing as pen or rubber or cylinder, but I cannot, by taking thought, discover it as paper or soft or cubical" (p. 52). Such a contrast is intended to illustrate the distinction between the given and its interpretation. "Pen", "rubber", and "cylinder" are all possible designations of this thing, contingent upon variable interest; as Lewis says, "My designation of this thing as 'pen' reflects my purpose to write; as 'cylinder' my desire to explain a problem in geometry or mechanics. . . . These divergent purposes are anticipatory of certain different future contingencies which are expected to accrue, in each case, partly as a result of my own action" (p. 52). By contrast, my inability, by taking thought, to discover this thing as paper, soft, or cubical represents a constraint on my apprehension or conceptualization which reflects the constancy of the given.

[7] Jerome S. Bruner, Jacqueline J. Goodnow, George A. Austin, *A Study of Thinking* (New York: John Wiley & Sons, Inc., 1956; reissued, Science Editions, Inc., 1962), p. 47.

[8] Martin Scheerer, "Cognitive Theory," chap. 3 in Gardner Lindzey, ed., *Handbook of Social Psychology* (Reading, Mass: Addison-Wesley Publishing Company, 1954), Vol. I, p. 110.

Note first that such constraint appears to conflict with Lewis' basic thesis that "The pure concept and the content of the given are mutually independent; neither limits the other" (p. 37). However, such conflict may be supposed merely apparent: the pure concept is unlimited in the sense that the activity of thought is free to frame whatever concepts it chooses; it is not equally free to apply them to the given however it may please to do so. In particular, the constraints on such conceptual application under variation of mental attitude provide an index of the given, and enable me to identify it and distinguish it from interpretive elements in my experience. That I cannot, by taking thought, apprehend this thing as paper is thus, for Lewis, a sign that its being non-paper is given to me.

Notice now, however, that the above reference to "taking thought" introduces a special condition that renders the criterion of the given, at the very least, ambiguous. That I cannot now, taking thought, imagine myself to have any interest which would lead me to call this thing "paper" does not show that *no* change in my attitudes, expectations, or conceptual equipment would, in fact, lead me to call it "paper". The given is supposed, for Lewis, to mark what would remain constant in my experience even if I were an infant, a member of a different culture with radically different upbringing, or, say, a subject in a psychological experiment designed to show the effects of motivation or belief on perception. Such constancy can certainly not be presumed to be the same constancy which survives my present attempts, by taking thought, to imagine circumstances of varied purpose in which my designations would be altered. These present attempts are, after all, themselves severely restricted by my present conceptual and attitudinal equipment. The fact that these attempts seem to yield any constancy at all in examples such as that of the pen is thus no indication whatever that there is a given in the original sense of something sensuous which "remains unaffected by any change of mental attitude or interest" (p. 66). The question can surely not be decided by thought experiments, and I have earlier argued that the psychological evidence weighs heavily against the supposition of such a given.

The reference to "taking thought" promises, at any rate, one clear advantage: I can presumably locate the given by means of an easily applicable criterion, dependent only upon my own powers of imagination. Unfortunately, as I have just argued, what is thus located turns out to be not the given at all, in the requisite original sense. Furthermore, the criterion itself is not nearly so clear as it may at first seem.

Presumably, the availability, for different purposes, of the alternative designations "pen", "rubber", and "cylinder" should lead me to regard them all as alterable and, hence, as belonging not to the given but to its interpretation. But how then can I take my inability to apprehend this thing as paper or soft or cubical as representing a constraint of the given? Surely, that this thing is non-paper, non-soft, and non-cubical cannot be given since the designations "non-paper", "non-soft", and "non-cubical" are equally recognizable as alternative labels suitable for different purposes. In fact, no designation whatever can represent the given, by the same token, for there are always alternative designations available for anything.

Perhaps, then, the crucial reason for affirming the givenness of this thing's being non-paper, non-soft, and non-cubical is just that I cannot bring myself to apply any of the *contrary* notions "paper", "soft", and "cubical" to what is before me, even under imagined variations of my present interests. Despite the fact, for example, that "non-paper" has alternative labels which I recognize as available (e.g., "non-soft", "pen"), these alternatives emphatically do not include the contrary designation "paper", in particular. But if inability to entertain the contrary is the critical mark of a designation reflecting the given, then "pen", "rubber", and "cylinder" must each belong also to the given rather than to its interpretation, for each satisfies the selfsame criterion of unalterability: I cannot, if I take this thing to be a pen, also apprehend it as a non-pen, nor can I, analogously, apprehend it as non-rubber or non-cylindrical, even under imagined variations of my present interests with regard to it. Whatever, in fact, I hold to be true of anything involves a constraint against my holding its contrary equally true of it; every designation, by this standard, turns out equally a reflection of the given. It seems, then, impossible to avoid the conclusion that no matter how Lewis' criterion be interpreted, it fails to yield the intended contrast between given and interpretation, on the basis of alterability. Every description is alterable in having alternatives, and unalterable in reflecting constraints. The establishment of an independent given cannot, I conclude, be achieved by appeal to some special sort of descriptive permanence.

But, you may say, Lewis holds the given to be ineffable, and hence untainted by the very possibility of error which all description introduces. In fact, you may argue, the independence of the given is essentially a matter of its freedom from error, that is to say, its certainty, and it is *certainty* that finally underlies Lewis' talk of the unalterability of the given. There is surely a substantial basis in Lewis' discussions for such an interpretation. Nevertheless, I believe that it cannot be formulated coherently, much less sustained by argument.

It should first be noted that the doctrine of the ineffability of the given seems to collide head-on with Lewis' illustrations of its unalterability. To illustrate the given by contrasting the variability of the designations "pen", "rubber", and "cylinder" with the alleged impossibility of apprehending the pen as soft, for example, seems clearly to imply that the hardness of the pen is given; it is also thereby to have provided a *description* of it as hard, or non-soft. Again, to illustrate the given by saying that "the pleasantness or fearfulness of a thing may be as un-get-overable as its brightness or loudness" (p. 57) is to have provided, in the same breath, four designations, "pleasant", "fearful", "bright", and "loud", which belie the ineffability of the given. Indeed, to speak of the given in any way at all, e.g., in an epistemological vein as an element of knowledge, is to purport to describe it. To describe it, furthermore, as ineffable is to cut the ground from under this very description itself. One cannot, in short, describe anything without describing it: if this be a trivial truth, so much the worse for any view which denies it, for it will be momentously false.

The notion of an error-free or certain given is, moreover, confused. Error and certainty, like truth and falsehood, are purported characteristics of descriptions, not in general of things described. If tables, for example, cannot be mistaken, this is no sign of their infallible truth, but rather a symptom of their ineligibility for either truth or falsehood. To speak of given qualities as incapable of being mistaken is, similarly, no evidence of their certainty, but rather a reflection of their being, like tables, non-descriptions.[9] If one persists in taking the given as certain because error-free, he will have to attribute certainty also to tables, stars, trees, and clouds; he will have effectively destroyed any philosophical point there might have been to the issue of certainty.

On the other hand, assuming that the notion of certainty is limited in its application to descriptions alone, then, since describability for Lewis always introduces the possibility of error, it turns out that if there are any descriptions of the given, none of them is, in actuality, certain. In sum, if the given is indeed ineffable, it makes no philosophical *sense* to call it certain, whereas if it is describable, none of its descriptions will in fact *be* certain. To put the matter another way, the sense in which the given is not subject to error is one which is entirely incapable of showing reports of the given to be immune from mistake. The so-called certainty of the given is thus of no epistemological interest whatever. The illusion of relevance is perhaps partially due to thinking of the given as "the brute-fact element in perception, illusion and dream" (p. 57), for facts are sometimes identified with things, sometimes with their descriptions; the inapplicability of error to the things themselves is thus illicitly transformed into the certainty of their descriptions. To recognize the ambiguity, however, is to be free of the illusion.

The consequences of these reflections are, I believe, far-reaching. The whole attempt to provide a solid mooring for our knowledge in an unalterable phenomenal base, along the lines of Lewis' *Mind and the World Order,* breaks down decisively. If the foregoing criticisms are correct, this attempt must be given up. It follows, further, that the interpretation of objectivity cannot properly be made to rest on it. In particular, the independent control which observation exercises over conception cannot be tied to the general project of

[9] See, in this connection, Nelson Goodman, "Sense and Certainty," *Philosophical Review,* LXI (1952), 161–162.

building knowledge on unalterable phenomenal foundations. This negative conclusion still leaves us with the task of offering a positive interpretation of observational control, but it liberates us from hampering conditions and points us, I believe, in a more promising direction.

It will be well to summarize our findings concerning the independence of observation to this point, so that we can assess our present situation before looking again at our main problem. We have, first of all, denied that observation is carried on in actual isolation from processes of conceptualization. We have, furthermore, denied that what is observed is unalterable by conceptual change. We have, thirdly, denied that what is given to observation is ineffable. Finally, we have denied that observational descriptions are certain, and we have charged with meaninglessness the notion of error-free sensuous content. If, then, our view of observational independence is to be consistent with these denials, it will need to accept the continuity and interaction of observation and conceptualization, and it will further need to acknowledge the fallibility of observational reports. Observational independence, in short, must require neither observational purity nor observational certainty. But can observation be independent in any relevant sense if it is both impure and fallible? That is our problem.

The first step we need to take is to break down the notion of conceptualization, for its unanalyzed use blurs a broad distinction critical for our topic. This distinction may be put in somewhat different ways, but the variations are irrelevant to the main issues at hand. We may express it, for example, as a distinction between concepts on the one hand and propositions on the other, between general terms or predicates on the one hand and statements on the other, between a vocabulary on the one hand and a body of assertions on the other, between categories or classes on the one hand and expectations or hypotheses as to category membership on the other.

A general term or concept applies to each of a number of things which may be said to belong to the class or category determined by it. To adopt a given term is, in effect, to commit oneself to recognizing anything satisfying this term; it is to adopt a certain scheme for the acknowledgment of instances.[10] "To learn 'apple'," says W. V. Quine, "it is not sufficient to learn how much of what goes on counts as apple; we must learn how much counts as *an* apple, and how much as another. Such terms possess built-in modes, however arbitrary, of dividing their reference."[11] A vocabulary of terms or concepts thus may be said to represent a means of delimiting and sorting what are taken to be things; it is a device of reference. Concepts or terms are not, however, in themselves capable of conveying propositions, beliefs, or assertions nor, hence, of making claims to evidential warrant or truth. The latter functions are dependent upon an available conceptual apparatus, to be sure, but they do not simply reduce to this apparatus; they represent rather a specialized accomplishment that may result from its proper use, going beyond mere reference, to the level of *statement*.

There is an ambiguity on this score which occasionally infects the notion of language. A language in the sense of a vocabulary and grammar is clearly different from a language in the sense of a body of assertions expressible by means of such vocabulary and grammar. Every language in the first sense makes possible the formulation of alternative and, indeed, conflicting systems of propositions or assertions; only languages in the second sense may be evaluated as to truth, consistency, or evidential warrant. The notion of conceptualization too is, unfortunately, afflicted with a parallel ambiguity, encompassing both senses of "language" just distinguished.

Putting the matter in terms of categories and simplifying, we may express the point as follows: Conceptualization relates both to the idea of categories for the sorting of items and to the idea of ex-

---

[10] Compare Nelson Goodman's comments on the determination of the realm of individuals to be recognized by a logical system: "This is commonly done in terms of the special primitive predicates chosen. . . . The individuals recognized include all that satisfy at least one place of at least one of these primitive predicates . . . ." *The Structure of Appearance* (Cambridge, Mass: Harvard University Press, 1951), p. 85; second edition (Indianapolis: The Bobbs-Merrill Co., Inc., 1966), ch. 3, sec. 11, p. 117.

[11] Willard Van Orman Quine, *Word and Object* (New York and London: The Technology Press of the Massachusetts Institute of Technology, and John Wiley & Sons, Inc., 1960), p. 91.

pectation, belief, or hypothesis as to how items will actually fit available categories; it links up with the notion of *category* and, also, with the quite different notion of *hypothesis*. The very same category system is, surely, compatible with alternative, and indeed conflicting hypotheses: that is, having adopted a given category system, our hypotheses as to the actual distribution of items within the several categories are not prejudged. Conversely, the same set of hypotheses may be formed compatibly with different category systems: the categories specifically referred to in these hypotheses may belong, as a common possession, to the different systems in question. Simply to set up an alphabetical filing system for correspondence is not yet to determine how tomorrow's correspondence will need to be filed. Conversely, to guess that the next letter will need to be filed under "E" or "L" is a prediction that may be made whether or not we have a place for "X" in our system.

A filing system may be said to determine in advance how correspondence is to be sorted. It does not determine in advance how any particular letter will be sorted. The second step in dealing with our problem is to separate these two notions of determination. A category system, within a limited context, may be described as imposing order in general and in advance on whatever experience in that context may bring. It commits us to ways of delimiting items to be recognized, as well as to modes of classifying them. Lacking such order altogether, we may, indeed, aptly be described as facing an undifferentiated chaos, since we lack the very recognition of things—that is to say, we do not individuate and separate items as objects of reference. Yet having a category system, we do not thereby prejudge the manner in which we shall need to apply it in the future. Without a vocabulary and grammar, we can describe nothing; having a vocabulary and grammar, our descriptions are not thereby determined.

Categorization does not, in other words, decide the forms of distribution which items will in fact display, nor does it, in itself, determine the categorial assignments of any particular item or class of items yet to be encountered. Such special anticipations may, however, be expressed by suitable hypotheses. Categorization provides the pigeonholes; hypothesis makes assignments to them. It is crucial to see that alternative assignments stipulated by different hypotheses are yet possible with the same categorization, for this provides at least the beginnings of an answer to our main problem. It means that we can understand a hypothesis which conflicts with our favored hypothesis of the moment, in terms of the very category system to which the latter appeals. In order for us to grasp such a conflicting hypothesis we do not need to decategorize and un-structure our thought, for such a conflicting hypothesis represents simply a different assignment under the selfsame category system. It follows that stripping away a given hypothesis is not tantamount to destroying our conceptual apparatus, so that we are left to cope only with a homogenized and ineffable given. On the contrary, it is compatible with retention of our category system, by reference to which conflicting assignments may be formulated and under-stood, and by means of which new data may continue to be processed, independently of our initial hypothesis.

To say that a categorization is independent of—in the sense that it does not prejudge—the particular hypotheses expressible through it means that observation may be conceived as thoroughly molded by such categorization and yet equally independent of these hypo-theses. We simply have a false dichotomy in the notion that obser-vation must be either a pure confrontation with an undifferentiated given, or else so conceptually contaminated that it must render circular any observational test of a hypothesis. Observation may be considered as shot through with categorization, while yet supporting a particular assignment which conflicts with our most cherished current hypothesis. It may be critically independent of such hypo-thesis while retaining its full categoricity, for categorization is itself, as we have seen, independent of any particular assignment of items to categories. We have here a fundamental source of control over the arbitrariness of belief.

What has thus far been said provides also an answer to the *paradox of categorization* mentioned in the previous lecture, to wit, if my categories determine what I observe, then what I observe fails to control my thought, whereas if my categories fail to determine what I observe, observation is impossible or useless. We need to recall, in this connection, the distinction between the general de-

termination effected by a category system, and the special determination in which a particular assignment to categories consists.
My categories may be said to provide a general determination of
fresh data: they reflect my advance resolution to individuate,
group, and separate such data along certain lines. They do not,
however, determine any particular assignment; in themselves they
do not compel me to choose one hypothesis rather than another.
My choice of specific hypotheses is thus, in particular, not prejudged by my categorization.

In the previous lecture I mentioned also the *paradox of common
observation,* according to which theoretical differences imply categorial differences and so, necessarily, observational differences.
At least one part of the answer to this paradox is already available.
For theoretical differences are, as we have seen, compatible with any
given categorization. It follows that opposing theorists are not
doomed to conflicting categorizations; such theorists may share the
same category system and indeed understand their conflicting hypotheses by reference to the common rubrics of this system. They are
not limited, of necessity, to the sharing of a formless and ineffable
given alone. Moreover, and this is a further point, even where their
categorizations are not, as wholes, identical, they may overlap with
one another, enabling not only commonly structured observations
but even (as we have earlier remarked) identical assignments,
through the very same hypotheses regarding certain classes of items.
Furthermore, there is no theoretical limit to the development of
overlapping categorizations where none existed before; were this
not so, we should be faced by a problem greater than the one posed
by our original paradox, that is, to explain how theoretical agreement could ever arise.

Finally, suppose two theorists with non-overlapping category
systems. Though they cannot even be opposed in the sense of being
able to express directly contradictory hypotheses in their two systems, we are not ourselves driven to describe them as necessarily
observing different things. Although we may find it plausible to
say that their systems determine observational differences, in the
sense that they correspond to different sets of readinesses to organize fresh experience, we are not thereby compelled to say that these

systems are directed to altogether different items. A category, as we have earlier urged, *both* delimits items to be recognized, *and* sorts them; the two functions of individuation and classification need not be construed as separated in actual fact or as performed by different mechanisms. Yet, the same form of individuation into separate things may be shared, over a given range, by two non-identical categories; compare, for example, "giraffe" and "animal." Categories may, generally, have in common certain identical items, although they group them differently. Two category systems, as wholes, may even have all their items in common, though classifying them in different ways. We may be inclined to say that these items are, correspondingly, observed in varying ways under the two categorizations in question; it does not follow that the items are not identical. An analogous point is expressed, in Quine's terms, as a distinction between ontology and ideology, and in Wittgenstein's terms, as a contrast between seeing x and seeing x as something or other.[12]

Nor does anything prevent us from trying to explain the *genesis* of categorizations differing wholly in *individuation:* we may characterize their divergent items as having been differently composed out of elements admitted by us, though not uniformly by them. Such an attempt does not bear directly on the paradox of common observation; however, it mitigates the isolation suggested by it insofar as variant categorizations, and attendant observational variations, are seen as sharing a common genetic background. Our explanatory scheme will certainly commit us further in both ontology and ideology, but such commitments are unavoidable, by our old principle that one can't describe anything without describing it. An epistemological theory, like any other, must take a stand on what there is, without suggesting either that there exists universal agreement on the question, or that there is only one possible stand to be taken on it, irrespective of problem and purpose. But an epistemological theory need not be any more arbitrary than any other

---

[12] Willard Van Orman Quine, *From A Logical Point of View* (Cambridge, Mass.: Harvard University Press, 1953), pp. 131–132; and Ludwig Wittgenstein, *Philosophical Investigations* (Oxford: Basil Blackwell, 1953), pp. 193ff.

theory, for its own hypotheses are themselves not prejudged by its categories, as we have earlier argued for the general case.

The main point advanced thus far is that hypotheses are controllable by observation even if observation be construed as fully categorized. This point is central, for hypotheses make truth-claims and, correspondingly, invoke the notion of a genuine test of such claims. Category systems as such, by contrast, do not qualify as being either true or false; in themselves they make no predictions and are therefore not subject, in the same way, to the test of observation. They are consonant with any distribution which the data may form in actuality, even one in which the data fall altogether outside the special rubrics into what is most conveniently thought of as the "miscellaneous" class.

Yet, certainly, category systems may be altered, for good reasons. Nothing said thus far should be taken to imply that our categorizations are themselves fixed. In practice, we frequently revise such categorizations, though not because we judge them false; not being assertions, they cannot be false. However, they may be judged insufficiently useful for the purposes at hand in light of the data distributions which they yield over a considerable time, distributions themselves describable under our initial categories. In effect, we take the test of certain hypotheses as relevant not only to a judgment of their truth but as an index of the usefulness of the category scheme as a whole. For example, the hypothesis that no single (non-miscellaneous) category in a given scheme will remain empty over a stipulated, sufficiently long period may, if it proves false, lead us to revise the scheme by removing the empty rubric, as we might in fact do with the "X" in our letter-filing system. *Acceptance* of the scheme may be said, then, to be based on the hope that no category (except for the miscellaneous one) will remain empty; yet, the *scheme itself* does not prejudge the satisfaction of this hope but rather allows perfectly well for the expression of its frustration as a certain sort of distribution, i.e., one with an empty rubric. Thus it is that we may hold our categorizations themselves subject to revision and, hence, to continuing controls.

In sum, category systems make no arbitrary truth-claims for they make no truth-claims at all; they are, moreover, revisable under circumstances specifiable by their own means. If it be argued that there is an element of arbitrariness in the free creation of wholly new category schemes to begin with, it must be admitted that the same sort of arbitrariness, if such it be, characterizes all intellectual creation; such arbitrariness is, however, not inconsistent with control.

Is the problem now completely solved? I think not. What has so far been shown, I believe, is that independent observational control over hypothesis is *possible* without appeal to an uncategorized given, since the sort of determination effected by categorization falls short of a question-begging support of favored hypotheses. Arguments to the contrary have been rebutted essentially by distinguishing categories from hypotheses, and contrasting the general ordering imposed by the former with the particular categorial assignments predicted by the latter; observation "determined by," "dependent on," or "filtered through" categories is thus quite conceivable as independent of any special hypothesis under test, expressible through reference to such categories. The general view thus advocated seems to me to preserve a tenable notion of the objectivity of observation, and to do so, moreover, without presupposing that the given is ineffable, uninfluenced by categorization, or reported by statements that are necessarily certain.

Still, a rebuttal of arguments against the very possibility of objectivity in observation leaves open the question whether this possibility is generally realizable in fact. You will no doubt remind me that the psychological considerations earlier cited provide strong empirical support not only for the importance of categorial schemata in observation, but also for the enormous influence of expectations and beliefs, that is, of particular *hypotheses*, upon observational mode and even upon content. Categorization itself may not completely determine any special assignment of items, but acceptance of a hypothesis in advance does, in fact, limit our perception of alternative assignments, expressible though they are under the category system in question. Accepting a hypothesis tends to restrict our view to certain of our categories, that is, to those which accord with the hypothesis itself. Expecting the next item to belong to a particular category, we tend generally to see it so. To be sure, the possibility that this item falls outside the category in question is expressible in terms of our category system, and is ascertainable, if realized, by observation which is molded by this system. Nevertheless, the possibility in question is the less likely to be seriously entertained in observation, the firmer our commitment to a hypothesis which denies it. So much seems indeed to be the lesson of psychology and everyday life.

The fact seems to me undeniable: Observational support for an assignment contrary to an accepted hypothesis needs to persist longer and fight harder for a hearing than observational data which accord with expectations.[13] Yet such contrary indications can and do make themselves felt. Our expectations strongly structure what we see, but do not wholly eliminate unexpected sights. To suppose that they do would be, absurdly, to deny the common phenomena of surprise, shock, and astonishment, as well as the reorientations of belief consequent upon them.

What is the upshot? There is here no evidence for a general incapacity to learn from contrary observations, no proof of a pre-established harmony between what we believe and what we see. In the disharmony between them, indeed, lies the source of observational control over belief. Our categorizations and expectations guide by orienting us selectively toward the future; they set us, in particular, to perceive in certain ways and not in others. Yet they do not blind us to the unforeseen. They allow us to recognize what fails to match anticipation, affording us the opportunity to improve our orientation in response to disharmony. The genius of science is to capitalize upon such disharmony for the sake of a systematic learning from experience.

[13] Striking illustrations may be found in J. S. Bruner and Leo Postman, "On the Perception of Incongruity: A Paradigm," *Journal of Personality* (1949), *18*, pp. 206–223.

3

Meaning and Objectivity

OBSERVATION, I have argued, is capable of providing independent control over belief even though it is channeled by categories, influenced by set, and describable only fallibly. For it is still capable of clashing with expectation and producing shock and unsettlement, forcing a new equilibration of conviction so as to preserve credibility. Observation is, thus, critically independent of hypothesis, even though it is itself unfixed and impure. Can we, however, say the same about *meaning?*

In fact, arguments from meaning threaten to undermine, from a new direction, the very sort of objectivity we have been urging. Our focus has been the concept of a category. A category, however, is the denotation or extension of a term, the meaning of which is said to be qualified by the family of meanings in which it dwells. In general, terms possess meanings not as isolated attachments, but rather as organic qualities which accrue in virtue of systematic function within a framework of linguistic use and intent. What a term means is not something that is magically linked to its sound or printed shape. It is discovered not by examination of such sound or shape but rather by looking to its role within a whole language system. The meaning of a category name is thus dependent upon the language in which it has a place. To alter this language is to alter its relative location, and so its very meaning.

It follows that if, as earlier maintained, the same categorization allows the expression of alternative hypotheses, these hypotheses will nevertheless confer alternative meanings on the categorization in question. If I can indeed formulate the denial of my hypothesis in terms of the categorization by which it is expressed, I cannot accept such denial without altering the very meaning of this categorization, for such acceptance effects a change in my language system. If I am really to understand an opposing theorist with the same categorization, I must, in fact, learn his whole language and, at least tentatively, accept his whole framework of belief, in order to appreciate the different meaning which such categorization has for him. The sharing of categories therefore does not imply the sharing of meanings, and without the sharing of meanings there can be no genuine exchange of understandings. To understand another is to share his language, and this means to share his thought-world, his outlook, expectations, and beliefs, for it is only in terms of the latter that his physical utterances can be rendered intelligible.

Such is the gist of a subjectivism that appeals to meaning. It may be so interpreted as even to allow common observation across the boundaries of belief, but it draws the line at allowing common meanings. What is commonly observed, it is maintained, will vary in significance, and the category under which it is observed will bear a varying interpretation, depending on the whole belief system to which it is referred. True communication from within one such system to another is hence impossible. The scientist is trapped in the web of his own meanings.

Now the standard objective view described earlier tries to avoid precisely this predicament by distinguishing the observational or experimental level of science from the theoretical level. Science is pictured as a two-tier structure, with observational statements at the bottom and theoretical statements rising above them. Observational language enables us to formulate the common facts neutral to theoretical controversies. It also allows us to express purported laws which can be directly tested by observational or experimental procedures. Both these functions are made possible by the presumed fact that observational terms have independent observational or experimental meaning. By contrast, theoretical terms do not have such meaning, but gain intelligibility solely through the role they are given in theoretical context. Theoretical statements accordingly neither formulate observable facts nor express generalizations directly testable by observation of their instances. However, they are subject to indirect evaluation by observational check of whatever observational consequences they help us to draw, in combination with other suitable statements. Differing theorists are thus not isolated in non-communicating theoretical cells. They can step outside their theories during the process of testing, and engage in a common check of their respective observational consequences through a shared observational language. They can agree, furthermore, to assess their own theories through reference to common observational findings expressible in such shared language. The theoretical level of science is thus built on, and controllable by, the common observational level.

This two-tier picture of science is presented persuasively in Ernest Nagel's *The Structure of Science:*

> Perhaps the most striking single feature setting off experimental laws from theories is that each "descriptive" (i.e., nonlogical) constant term in the former, but in general not each such term in the latter, is associated with at least one overt procedure for predicating the term of some observationally identifiable trait when certain specified circumstances are realized. The procedure associated with a term in an experimental law thus fixes a definite, even if only a partial, meaning for the term. In consequence, an experimental law, unlike a theoretical statement, invariably possesses a determinate empirical content which in principle can always be controlled by observational evidence obtained by those procedures. . . .
>
> In contrast with what holds uniformly for the descriptive terms in experimental laws, the meanings of many if not all descriptive terms occurring in theories are not specified by such overt experimental procedures. To be sure, theories are frequently constructed in analogy to some familiar materials, so that most theoretical terms are associated with conceptions and images derived from the generating analogies. Nevertheless, the operative meanings of most

48   SCIENCE AND SUBJECTIVITY

theoretical terms either are defined only implicitly by the theoretical postulates into which the terms enter, or are fixed only indirectly in the light of the eventual uses to which the theory may be put. . . .

Accordingly, even when an experimental law is explained by a given theory and is thus incorporated into the framework of the latter's ideas . . . , two characteristics continue to hold for the law. It retains a meaning that can be formulated independently of the theory; and it is based on observational evidence that may enable the law to survive the eventual demise of the theory. . . . Despite what appears to be the complete absorption of an experimental law into a given theory, so that the special technical language of the theory may even be employed in stating the law, the law must be intelligible (and must be capable of being established) without reference to the meanings associated with it because of its being explained by that theory. . . . the terms must have determinate senses statable (albeit only partially) independently of the particular theory adopted for explaining the law.[1]

It should be clear why it appears advantageous to represent experimental laws as retaining at least a minimal independence of meaning. For, assuming such independence, we may plausibly say of these laws that they provide the material which theories strive to explain, and thereby serve as touchstones by which the relative explanatory success of theories is judged. "Indeed," as Nagel writes, "were this not the case for the laws which a given theory purportedly explains, there would be nothing for the theory to explain."[2] To interpret the experimental laws solely by means of the theoretical counterparts of their descriptive terms would remove them from the realm of observational ascertainability and would, moreover, eliminate the basis for assessing the instrumental worth of relevant alternative theories as explanations. On the other hand, upholding the meaning-independence of experimental laws, we can understand not only the process of theoretical explanation, and the analogous process by which one law is reduced to another, but also the relative permanence of experimental laws throughout changes of theory. To use Nagel's illustration, though Millikan's oil-drop

[1] Ernest Nagel, *The Structure of Science* (Indianapolis: Hackett Publishing Co., 1979), pp. 83–87.
[2] *Ibid.*, p. 87.

experiment was suggested by electron theory, the theory has been considerably modified since the experiment was first performed. Yet "the truth of the experimental law that Millikan helped to establish (namely, that all electric charges are integral multiples of a certain elementary charge) is not contingent upon the fate of the theory; and, provided the direct observational evidence for it continues to confirm the law, it may outlive a long series of theories that may be accepted in the future as explanations for it."[3]

Although the advantages of the two-tier conception of science are thus patent, the critical question is whether it can be sustained. How indeed are we to imagine the incorporation of an experimental law into a new theoretical framework—in fact, as Nagel puts it, its "complete absorption" into the theory "so that the special technical language of the theory may even be employed in stating the law"—and at the same time uphold the meaning-independence of the law? If complete absorption of the law into the theory takes place, so that it is literally restated in theoretical terms and treated as a derivation from theoretical postulates, how can the law be even partially independent of the theory in meaning? Conversely, if the terms in the experimental law must, as Nagel puts it, "have determinate senses statable (albeit only partially) independently of the particular theory adopted for explaining the law," how can the law itself be said to be explained by the theory, rather than merely some similar sounding string of noises or marks with different meaning, i.e., some sentence-sequence of homonyms? The meaning-independence of laws thus seems to conflict with their complete absorption into explanatory theories.

In the face of this conflict, some theorists have indeed, as we earlier noted, tended to surrender altogether the meaning-independence of laws. Thus T. S. Kuhn, as we saw, propounded the notion of "the scientific revolution as a displacement of the conceptual network through which scientists view the world"[4]—a displacement presumably not limited to the upper levels of science

[3] *Ibid.,* p. 88.

[4] Thomas S. Kuhn, *The Structure of Scientific Revolutions,* p. 101.

only, but penetrating also the very apparatus by which experimental facts and observational evidence are formulated. And P. K. Feyerabend has written:

> What happens when transition is made from a restricted theory T' to a wider theory T (which is capable of covering all the phenomena which have been covered by T') is something much more radical than incorporation of the *unchanged* theory T' into the wider context of T. What happens is rather a *complete replacement* of the ontology of T' by the ontology of T, and a corresponding change in the meanings of all descriptive terms of T' (provided these terms are still employed).[5]

His argument allows that "the principles of the context of which T' is a part need not be explicitly formulated, and as a matter of fact they rarely are." It is sufficient to suppose "that they govern the use of the main terms of T'." Given then that such principles are shown to conflict with T, "the idiom of T' must be given up and must be replaced by the idiom of T."[6] Furthermore, Feyerabend denies any special status for so-called observational language, holding that "even everyday languages, like languages of highly theoretical systems, have been introduced in order to give expression to some theory or point of view, and they therefore contain a well-developed and sometimes very abstract ontology." Under conflict "with newly introduced theories . . . they must therefore be either abandoned and replaced by the language of the new and better theories even in the most common situations, or they must be completely separated from these theories. . . ."[7]

---

[5] P. K. Feyerabend, "Explanation, Reduction, and Empiricism," in Herbert Feigl and Grover Maxwell, eds., *Minnesota Studies in the Philosophy of Science,* Vol. III (Minneapolis: University of Minnesota Press, 1962), p. 59.

An extended discussion of Feyerabend's views is contained in the papers of J. J. C. Smart, Wilfrid Sellars, Hilary Putnam, and P. K. Feyerabend, in Robert S. Cohen and Marx W. Wartofsky, eds., *Boston Studies in the Philosophy of Science,* Vol. II (New York: Humanities Press, 1965), pp. 157–261. An extended critical account of Feyerabend's views is given by Dudley Shapere in his recent essay "Meaning and Scientific Change," in Robert G. Colodny, ed., *Mind and Cosmos* (Pittsburgh: University of Pittsburgh Press, 1966), pp. 41–85.

[6] "Explanation, Reduction, and Empiricism," p. 75.

[7] *Ibid.,* p. 76.

It is worth noting one major puzzle in the first of the passages from Feyerabend just cited. He describes T as a theory wider than T' and capable of covering all the phenomena covered by the latter. Yet there is presumably no neutral way, on his account, to formulate the phenomena covered by both, the phenomena being determined by the available descriptive terms which have undergone crucial changes of meaning in every case. Indeed, earlier in the same discussion, he has urged that "Experimental evidence does not consist of facts pure and simple, but of facts analyzed, modeled, and manufactured according to some theory,"[8] arguing (especially from considerations relating to measurement) that "Hence, the 'facts,' within D', which count as evidence for T will be different from the 'facts,' within D', which counted as evidence for T' when the latter theory was first introduced."[9] Even where measurements converge, critical meanings differ, moreover, under theoretical change; in this vein, Feyerabend writes:

> Of course, the values obtained on measurement of the classical mass and of the relativistic mass will agree in the domain D', in which the classical concepts were first found to be useful. This does not mean that what is measured is the same in both cases: what is measured in the classical case is an *intrinsic* property of the system under consideration; what is measured in the case of relativity is a *relation* between the system and certain characteristics of D'. It is also impossible to define the exact classical concepts in relativistic terms or to relate them with the help of an empirical generalization. . . . It is therefore . . . necessary to abandon completely the classical conceptual scheme once the theory of relativity has been introduced; and this means that it is imperative to use relativity in the theoretical considerations put forth for the explanation of a certain phenomenon as well as in the observation language in which tests for these considerations are to be formulated. . . .[10]

If the facts, measurements, and observations themselves receive different formulations owing to the conceptual effects of theoretical change, what point can there be in Feyerabend's reference to phenomena common to differing theories, or even in his reference to the common domain D'? The question receives no clear answer in the

[8]*Ibid.*, pp. 50–51.   [9]*Ibid.*, p. 51.
[10]*Ibid.*, pp. 80-81.

discussion we have quoted, and it may be suggested that if these problematic references are not merely to be explained away as a *façon de parler,* they are indeed inconsistent with the general position put forth. But Feyerabend is, in any event, perfectly explicit on the main point of present interest to us: the alteration of descriptive meanings as a result of transplantation to a new theoretical context. The meaning-independence of laws in particular is, then, to be given up, their absorption into new frameworks of theory to be acknowledged as reconstituting the meanings of their constituent terms.

Such a course seems, however, to put us back on the road to subjectivism, for which the two tiers of science are, in effect, collapsed into one. Gone is the notion of a shared observational language with constant meaning, capable of controlling choices of theory. The very distinction between observational and theoretical language is erased, for learning a new theory is construed as acquiring new habits of "observational" description, while all purportedly pure observational language is seen as qualified by the "theoretical" belief-context in which it functions. No neutral observational description is available for those phenomena to be explained by theory, for there simply are no neutral observational descriptions. There is no real permanence of experimental laws under theoretical change, for such change alters the very meaning of the terms in the laws, thereby changing their significant identities as laws. It seems further to follow that there can be no genuine accumulation of empirical knowledge, no reduction of one law to another, and no explanation of a prior law by a theory succeeding it. At best, we can have the replacement of one general viewpoint by another. To converse with an opposing theorist on neutral ground is impossible; one must enter his belief system, however tentatively, in order to grasp his meaning. Conversation, in short, requires conversion. The result is what I called, in my first lecture, the *paradox of common language,* the isolation of each scientist within the world of meanings created by his own beliefs, so that every attempt to raise his voice to be heard beyond that world must fail, and every attempt to justify these very beliefs must itself presuppose them. Justification of theory in fact disappears, giving way to proclamation of theory.

As already remarked in the previous lecture, I believe this line of argument to be self-refuting. For if it is correct, it cannot hope to persuade those who are not already persuaded. Conversely, the very effort to convince opponents of the doctrine conflicts with the doctrine itself. Taking the doctrine at its face value, we have every right to refuse to accept it, on the ground that, imprisoned as we are in our own world of meanings, we cannot grasp its siren message from worlds beyond. However, we must do more than simply produce such a polemical reply. For *we do not* accept the doctrine in the first place, and cannot, in our own minds, claim that we do not understand it. In fact, we not only understand it, but find the considerations brought forward in its behalf to be initially plausible, irrespective of whether or not its proponents refute themselves in adducing these, or indeed, any other considerations whatever. The problem is thus not simply to show the doctrine self-refuting or otherwise defective. It is to show how we can ourselves reasonably avoid being driven to it.

As a preliminary to my own effort to deal with this problem in the following discussion, I shall here summarize briefly the main strengths and difficulties of the rival views we have been considering. The standard, two-tier view has the great virtue of accounting for theoretical explanation of laws previously accredited, and accounting also for the justification of such explanation, in particular cases, as a public and communicable enterprise. This achievement is made possible by the separation of observational from theoretical language, and the notion that experimental or observational laws retain at least some shareable constancies of meaning throughout changes of theoretical context. The difficult problems confronting this view, are, first, to render plausible the distinction between observational and theoretical language and, secondly, to reconcile the meaning-independence of laws with their complete absorption into alternative theoretical frameworks.

The contrasting one-tier view we have considered seems to be supported by the insight that meaning is not an atomic appendage to a term, but rather qualifies the term through its functional position within a whole framework of language. This view seems strong also in its criticism of the idea of a fixed and general division between observational and theoretical language, for surely this division is too vague and fluid to be taken as a constant linguistic fact. The main difficulties of the one-tier view arise, on the other hand, from the denial of meaning-independence to observational terms and laws. Theoretical explanation of prior experimental laws turns out to have no object; reduction and, what is more important, accumulation of laws are myths, the relative evaluation of theories by reference to common experience is impossible, and even the communication of argument across belief barriers breaks down. Is there, then, a way of avoiding the severe defects of both views, and incorporating their combined strengths into a new and coherent objectivism? I believe the answer is yes, and in the remainder of this lecture shall outline the reasons for this confidence.

Let me begin by attacking the notion of meaning. It is ironic in the extreme that this central notion, which is linked to intelligibility, communication, understanding, and significance, is itself wrapped in philosophical obscurity. We will do well to disentangle several of its strands as we reconsider the basic subjectivist argument with which we started. This argument, it will be recalled, stressed the dependence of meanings upon linguistic context, concluding that to understand another requires a sharing of the thought-world, that is, the framework of beliefs, embodied in his language. Now we must surely agree, and indeed insist, that whatever meanings may conceivably be, they cannot be construed as inherent in the physical constitution of the words with which they are associated in given contexts. The same word, considered purely as a physical pattern, may mean something different in different circumstances. It may, for example, undergo historical changes in meaning, take on varied nuances in various practical contexts, change characteristically with the speaker, occur with differing import in different languages, or be subjected to variant stipulations for varied technical purposes. The crucial factor is not its physical character but rather its role in the particular human context of its employment: what is intended— that is, meant—by its user in the use of it.

Here, however, lurks an important ambiguity, between the *sense* and the *reference* of an expression. Is the question of what is *meant* by the user a matter of the concept or idea *expressed,* that is, a question concerning the connotation, intension, attribute, or sense associated with the word? Or, on the other hand, is it rather a matter of the thing *referred* to, that is, a question of the denotation, extension, application, or reference of the word? *Both* sense and reference depend on the human context of word use; *neither* is magically inherent in the word's physical constitution. Both are, so to speak, elements not of nature but of convention. Yet they are different, and it is crucial for philosophical purposes not to confuse them.

Reference is surely the clearer of the two. The existence of things referred to by at least some of our words is an unavoidable presupposition of any view we might adopt; it is not a matter of controversy. The existence of senses or meanings, as entities underlying these words, is, however, not an indispensable assumption of any view, and it does not, in fact, command universal agreement. W. V. Quine has expressed the main points as follows (using the word "meaning" as "substantial meaning"):

> Confusion of meaning with reference has encouraged a tendency to take the notion of meaning for granted. It is felt that the meaning of the word 'man' is as tangible as our neighbor and that the meaning of the phrase 'Evening Star' is as clear as the star in the sky. And it is felt that to question or repudiate the notion of meaning is to suppose a world in which there is just language and nothing for language to refer to. Actually we can acknowledge a worldful of objects, and let our singular and general terms refer to those objects in their several ways to our hearts' content, without ever taking up the topic of meaning.
>
> An object referred to, named by a singular term or denoted by a general term, can be anything under the sun. Meanings, however, purport to be entities of a special sort: the meaning of an expression is the idea expressed. Now there is considerable agreement among modern linguists that the idea of an idea, the idea of the mental counterpart of a linguistic form, is worse than worthless for linguistic science.[11]

Dispensing with substantive meaning-entities does not require us to deny the underlying properties and distinctions which they are naively thought to explain. Without hypostatizing meanings, we are

[11] Willard Van Orman Quine, *From a Logical Point of View* (Cambridge, Mass.: Harvard University Press, 1953), pp. 47–48.

still free to describe individual expressions as meaningful and to judge pairs of expressions as identical in meaning. As Quine puts it, "The useful ways in which people ordinarily talk or seem to talk about meanings boil down to two: the *having* of meanings, which is significance, and *sameness* of meaning, or synonymy. What is called *giving* the meaning of an utterance is simply the uttering of a synonym, couched, ordinarily, in clearer language than the original. If we are allergic to meanings as such, we can speak directly of utterances as significant or insignificant, and as synonymous or heteronymous one with another."[12] Still the latter four adjectives, though not committing us to substantial meanings, are far from having generally accepted analyses or paradigms, or even generally agreed-on applications, as has been argued, notably by Quine, to the relative advantage of notions of reference.

Aside from the contrast in point of clarity, it is important for us to note that sense or meaning, even under adjectival construction, differs characteristically from reference. In particular, sameness of sense presupposes, but is not itself presupposed by, sameness of reference. Scott (to use Russell's example[13]) is the author of *Waverley;* the name "Scott" denotes the same entity as is denoted by the descriptive phrase "the author of Waverley," yet the name and the descriptive phrase surely differ in meaning. The expression "Evening Star" (to use Frege's example[14]) refers to the very same thing to which the expression "Morning Star" refers, and yet these expressions fail of synonymy; the discovery of their identical reference was astronomical, not linguistic.

It is, furthermore, of great importance to realize that we can, and indeed typically do decide issues of reference in specific cases without a prior characterization of sense or meaning in terms understandable to us. We confidently use the word "chair", for example,

[12] *Ibid.,* pp. 11–12.

[13] Bertrand Russell, *Introduction to Mathematical Philosophy* (London: George Allen and Unwin, Ltd., 1919), chap. 16.

[14] Gottlob Frege, "On Sense and Nominatum," tr. Herbert Feigl, in H. Feigl and Wilfrid Sellars, eds., *Readings in Philosophical Analysis* (New York: Appleton-Century-Crofts, Inc., 1949), pp. 85–102. Originally appeared as "Ueber Sinn und Bedeutung," *Zeitschr. f. Philos. und Philos. Kritik, 100* (1892), pp. 25–50.

in application to particular items in our environment, without first deciding on an acceptable synonym or synonymous defining expression for this word.

Indeed, were it not possible to gain an understanding of words in application independently of a prior elaboration of their meanings in understandable terms, we should be at a loss to explain how such understanding is acquired in the first instance. Were the child never to understand any word before grasping some synonymous defining expression, he could never begin to understand any word at all. For each such defining expression would itself require a further synonymous expression in order to be understood, and, barring unilluminating circularity and infinite regress in the chain of synonymies, the initial expression would, in every such chain, remain beyond his reach; the process could, in short, never get started.

We should note, finally, that, for the purposes of mathematics and science, it is sameness of reference that is of interest rather than synonymy, in accordance with the general principle that a truth about any object is equally true of it no matter how the object is designated. The identity of the Morning Star with the Evening Star, for example, is scientifically interesting for it allows us to interchange the two phrases "Morning Star" and "Evening Star" in any truth hitherto established with the help of either, thus extending such truth from the one to the other. The difference in meaning of these phrases is clearly no bar to such advantage. It concerns at best only the special subject-matter interests of the linguist and, as we have seen, poses even for him a knotty problem of clarification.

The concept of definition itself, in mathematical and scientific contexts, hinges not on synonymies but, at most, on referential equivalences. Speaking of the theory of definability, Quine writes that it

> stands clear of the theory of meaning altogether and falls square
> within the theory of reference. The word 'definition' has inde
> commonly connoted synonymy, which belongs to the theory
> meaning; the mathematical literature on definability, however, I
> to do with definability only in the following more innocuous se
> A general term t is said to be *definable* in any portion of langu
> which includes a sentence S such that S has the variable 'x' in it

is fulfilled by all and only those values of '*x*' of which *t* is true. Definability so construed rests only on sameness of reference—sameness of extension on the part of *t* and *S*. Definability of expressions of other categories than that of general terms may be explained in fairly parallel fashion.[15]

Goodman has, moreover, argued that an even weaker notion of referential isomorphism is sufficient for the definitions of explanatory systems of science and philosophy.[16] Indeed, the classification of any scientific proposition as definition or as empirical truth is thus largely a matter of choice in the interests of convenient systematic presentation for the purposes at hand; nothing beyond referential relations need, in any event, be taken into account. As for deduction within scientific systems, it should be especially noted that it requires stability of meaning only in the sense of stability of reference in order to proceed without mishap. That is to say, alterations of meaning in a valid deduction that leave the referential values of constants intact are irrelevant to its truth-preserving character.

With these several points in mind, let us return to the basic subjectivist argument from meaning with which we have been concerned. A category term, it was said, derives its meaning from its role within a language system; its meaning is not an atomic somewhat, mysteriously linked to its physical character. But then to alter the language system in any way is to alter the meaning of the category term. Thus, it was concluded, to share such meaning with another is to share all his substantive beliefs as well.

Now it is clear upon reflection that the alternatives offered us by this argument are far from exhaustive: either meaning is some queer sort of metaphysical appendage to sounds and shapes, or else it is related in one-to-one manner with whole language systems, inclusive of substantive beliefs. Certainly, meaning is not, as we have already stressed, a function simply of the physical constitution of terms; it depends also on the human context within which they are used. It does not at all follow, however, that it must therefore have a one-to-one relationship with language systems, systems understood, moreover, not simply as embracing distinctive statement-

---

[15]*From A Logical Point of View*, p. 132.
[16]Nelson Goodman, *The Structure of Appearance*, chap. 1.

forming mechanisms, but as expressing distinctive bodies of belief.

Even proponents of meaning as sense or connotation operate with this notion, in effect, only within the bounds of synonymy determinations. Typically, they have imagined such meaning to flow directly from the apparatus of a language, considered as including or authorizing certain assertions with an official status respecting meaning. A language system, under such a conception, embraces some device designating officially synonymous expression-pairs or specifying directly a set of synonymy-based, or so-called analytic, truths. However, such a system does not further provide an official choice among assertions expressible within its notation but independent of synonymy considerations, that is, it does not select "synthetic" or "empirical" truths. Language systems are, from the point of view of a concern with meaning, no more to be equated with descriptive or empirical axiom systems, than dictionaries of French are to be confused with textbooks of chemistry written in French. Such a confusion, however, underlies the argument from meaning we have been considering. For this argument supposes meaning to be dependent upon the language system, construed not simply as encompassing what is derivative from synonymy considerations, but as including also all hypotheses or assertions taken as true—in short, the whole framework of accepted beliefs. Only thus can the argument proceed to the conclusion that shared meaning presupposes shared belief. Once the notion of a language system in the restricted sense is clearly differentiated from that of an axiom system generally, the conclusion no longer follows, even though meanings are still construed as relative to the synonymies specified by a given language system.

Nor are these meanings themselves so interlocked that a change in any one affects all the rest within a given language system. Given any particular term, the synonymy relations of which are fixed relative to a designated system, it by no means follows that such relations cannot persist throughout synonymy changes affecting other terms within the system in question. Meanings, that is, may be considered as *relative to* language systems without compelling the further assumption that they cannot be *shared by* different language systems. Since each such system is, moreover, itself capable of being embedded within an indefinite variety of descriptive axiom systems, the synonymy network (or meaning) of any term may persist throughout indefinitely wide variation of substantive belief. In sum, though connotative meaning is relative to language, it may remain fixed throughout alterations of belief; opposing theorists may employ selected synonymous terms, and indeed the same language system, though holding conflicting beliefs.

The latter conclusion is strengthened if we now turn from connotation altogether and think of meaning not as sense but as reference. For identity of reference, as we have earlier noted, does not imply sameness of sense. Terms may denote the very same things though their synonymy relations are catalogued differently. Hence common reference may not only survive alterations of belief outside the realm of synonymies and variations affecting the synonymies of neighboring terms within a given language system; it may also survive synonymy alterations bearing directly on the very terms in question. Opposing theorists may differ in respect of these latter alterations, or may reject the idea of specifying synonymy relations altogether, and they may yet mean, that is, refer to, the same things. Their differing beliefs will then represent conflicting systems of assertions expressible with the same language apparatus, the latter individuated simply by its logic and its referential properties. Even, moreover, if their languages differ as wholes, they may well overlap with respect to certain terms, to which they assign identical referential functions, and by means of which their opposing theories may be expressed.

In any event, it will still be the case that meaning as we are now considering it, i.e., reference, is relative to language though shareable by theoretical opponents, and is decidedly not an inherent concomitant of the physical constitution of terms. The fundamental subjectivist argument from meaning thus collapses. If meaning is not a mystery of sounds and shapes, neither is it a subjective prison of the mind. It is perfectly possible to share common meanings with those who disagree with us in belief. Since, indeed, disagreement, in the sense of explicit contradiction, requires common meanings in order to be differentiated from a mere changing of the subject, the collapse of the subjectivist argument reinstates the possibility of such disagreement, itself involved in any plausible conception of rational discussion.

The collapse of the subjectivist argument does not in itself, however, resolve all the issues that have here concerned us. We still need to deal, in particular, with the opposition between the one-tier and the two-tier conceptions of science, described earlier against the background of the subjectivist argument, but requiring independent consideration. The two-tier conception, you will recall, rests on a basic contrast between observational and theoretical language, which is presumed to guarantee a minimal meaning-independence for experimental laws. This conception faces, however, the problem of reconciling such independence with the absorption of laws into varying theoretical frameworks. By contrast, the one-tier view denies meaning-independence to experimental laws and indeed rejects the notion of a fixed distinction between observational and theoretical language. However, it pays the price of denying also the full explanatory role and justification of theories, as well as the reduction and accumulation of experimental laws.

The main proposal I should like to make here with respect to this opposition of views is that we reconstrue the possible meaning-independence of experimental laws, and indeed of observational formulations generally, in terms of reference rather than sense or connotation. Experimental or observational laws, for example, formulate relationships of one or another sort among the classes of elements denoted by their constituent terms. The denotations or references of these terms in specific cases can, as we have earlier stressed, be determined independently of a characterization of their respective senses. Nor will varying assignments of sense to these terms, if compatible in every case with their determined denotations, necessarily disrupt deduction; for their actual denotations, as wholes, may well be assumed to remain unchanged through such variation, identity of actual reference being, in any event, consistent with alterations of sense. Moreover, aside from the question of varying sense or connotation, there is surely no doubt that a statement with fixed referential interpretation may be validly inferred from different sets of premisses, such interpretation itself resting on determinations independent of prior agreement on these premisses.

Now it follows from such a view of the matter that the absorption of experimental or observational laws into different theoretical frameworks is compatible with their constancy of referential interpretation. Alterations of theoretical framework, though they be understood to effect changes in relevant substantive premisses, in official definitions, and even in the senses of constituent terms, need not also be taken thereby to alter the referential meaning-identity of critical experimental formulations. Such identity may, of course, be deliberately changed for independent reasons. But the mere fact of absorption into varying frameworks of theory does not, in itself, require us to say that the old laws have altogether changed, giving way to new. The constancy of referential interpretation is, moreover, accessible to reinforcement through shared processes of agreement on particular cases; I suggest that such accessibility is, indeed, the core of the idea that experimental laws are couched in a relatively independent observational language. To this question we shall, however, return later in this lecture. Here the main point to be stressed is that laws may retain their referential identities throughout variations of theoretical context.

The possibility of such constancy is sufficient, I believe, to give body to the notion of theoretical explanation of experimental laws and of observational formulations generally, thereby lending substance also to the notion of control or justification of theory. Such constancy also renders plausible the possibility of the reduction of experimental laws and the accumulation of such laws, though it does not, of course, require that what are now taken to be laws will in fact, or ought in principle, to be retained in any future corpus of science. The insistence of the one-tier view that theoretical incorporation affects meanings is plausible at best only with respect to senses, and even so only for certain theoretical incorporations, i.e., those considered as altering synonymy-relationships in some way or other. Such alteration, however, in the first place, does not automatically effect a disruption of referential constancy nor, therefore, does it automatically disturb the deductive relations which underlie reduction and explanation. In the second place, whether synonymies are sustained or violated is, in itself, as I have previously urged, of no real interest for the general purposes of science.

My account of the meaning-independence of experimental laws in terms of reference seems to me to provide an answer to the outstanding problem of the two-tier conception, that is, how to reconcile such independence with varying theoretical interpretation. This account thus enables us to preserve the advantages of the two-tier view as outlined in Nagel's exposition, cited earlier; sympathetic to his intent and purpose, it seems, moreover, to provide an answer to the problem noted in connection with his formulation: he describes the meaning-independence of experimental constants in terms of independent senses arising from their association with experimental procedures. The law, in his view "retains a meaning that can be formulated independently of the theory."[17] The terms in experimental laws "must," he writes, "have determinate senses statable (albeit only partially) independently of the particular theory adopted for explaining the law." Additionally, he speaks of the law as having "meanings associated with it because of its being explained by that theory."[18]

The problem is that we are presumably being asked to imagine a statement, under one assignment of senses to its terms, to be explained by a theory which makes a different assignment of senses to these same terms. If meaning is, in effect, interpreted as sense, how can we avoid individuating statements by their sense assignments, and how can we then avoid the conclusion that experimental laws, so-called, are really ambiguous statements, constructed out of homonymous constants? Why, indeed, should the explanation of a law, individuated in terms of one determinate meaning, carry over to the explanation of the same words when given a different determinate meaning and thus individuating a different statement? Surely, the mere fact that the same words, as physical tokens, are held in common is insufficient to support such an explanatory extension. In short, if the terms of an experimental law have senses independent of those attaching to its theoretical restatements, these independent senses cannot also, without fatal ambiguity, characterize the varying restatements themselves. The meaning of the law-state-

[17] *The Structure of Science,* p. 86.

[18] *Ibid.,* p. 87.

ment thus fails, and must fail, to persist throughout variable theoretical restatement.

By contrast, the version of meaning-independence which hinges solely on reference is free of this difficulty, for identity of reference is altogether compatible with variable sense. Individuated by their referential values rather than their sense assignments, laws may be interpreted as retaining their full identities under changes of explanatory framework. We may, further, rest assured that homonymies of sense which do not destroy referential individuation are powerless to disrupt the deductive relationships basic to explanation, reduction, and systematization. It should once again be urged, therefore, that presumed alterations of sense are of no general scientific interest anyway, deductive systematizations in terms of new theories being prized irrespective of their effects on alleged synonymies of constituent terms.

I have thus far sided with the two-tier conception and rejected the main contention of the one-tier view. But there is one important issue, that concerning observationality, on which I believe the latter view to have the advantage. Proponents of the two-tier conception have sometimes spoken as if the division between observational and theoretical language were general and constant, though vague at the boundary. Such a doctrine is, I believe, vulnerable to the criticism that the boundary is not only subject to varying judgments, but that it is, in any event, itself in motion: new forms of description must be acknowledged to arise both naturally and by invention, and theoretical development, in particular, often brings to the fore novel modes of evidential formulation employed in the assessment of further theory. There is more involved here, it would seem, than vagueness at a fixed borderline between observational and theoretical language.

What is important in this context, as earlier suggested, is the possibility of shared processes of decision on the referential force of a term by application to cases. Such possibility allows for an independent understanding of the referential interpretation of a law by opposing theorists, and is thus sufficient to prevent the subjectivist dénouement. It is, however, compatible with this possibility to allow that what is judged decidable by application to cases may well vary with history, purpose, and prior theoretical context. Such variation is an index of the fact that referential modes are not forever fixed, that the learning of new languages of reference is possible. The relative independence of observation from theory must not be taken to imply that there is some single descriptive language, fixed for all time, within which science must forever fit its experimental accounts of nature. Nor, surely, should one imagine these accounts to accumulate within the scope of such a language by some iron law of progress; as we have noted above, there is no guarantee that what are now thought to be laws will be retained under the impact of further experience, even when the descriptive scheme remains constant. Such accumulation as in fact occurs need, moreover, not be construed as wholly contained within a unique descriptive language, fixed in advance. It is open to us to acknowledge a multiplicity of schemes of reference, shared in part or in whole, overlapping in lesser or greater degree, and suitable in varying ways for different purposes, yet allowing, within their several ranges, for the growth of experimental knowledge in a controlled manner.

Given any such scheme, the contrast between observation and theory marks the fact that in our relatively independent judgment of cases within this scheme, we have a device of control over systematization and general conception. Such control does not determine, to be sure, but constitutes an independent factor in the continuing accommodation of our judgments. An analogy may perhaps be suggested in the interplay between case and code in the sphere of the law, or in the dynamic relation between our particular moral reactions and our general principles and conceptions of morality.

In the matter of the critical reception of newly introduced theories, the reference to observation has moreover, I believe, a special point.[19] A new theory arising within a given referential tradition cannot command initial consensus on presumably confirming cases of its own, but must prove itself against the background of prior judgments of particulars. It must acknowledge the indirect control of accumulated laws and theories encompassing already crystallized judgments of cases. Even if some such judgments are to be chal-

---

[19]With respect to the topics discussed in the present paragraph, I have benefited much from conversations with Sidney Morgenbesser.

lenged from the start, the challenge needs to be expressed in a form that is intelligible for the received descriptive mode, and special motivation for the challenge must, of course, be adduced. The inventor of a new theory cannot, at the outset, motivate his new forms of discourse by simply saying "Look and see!" Now, it may turn out that this new theory, having won an initial place largely through indirect forms of argument against the background of acknowledged facts, eventually forces a revision of older judgments of cases and, what is more significant, perhaps, opens up new ranges of evidential description, thereafter developing consensus on relevant instances of its own. It remains true that, at the outset, its advantage needs to be shown in the context of judgments of cases already available, and in relation to the scheme by which such judgments are formulated. Despite the varying perspectives from which such judgments are made, they thus provide a genuine measure of control over new theory at every juncture. Such control, persistent through changing perspectives, is a characteristic mark of science.

4
Change and Objectivity

WE HAVE already removed much of the sting of subjectivist arguments appealing to the facts of change. For we have denied that objectivity requires either a fixed given or immutable descriptions thereof. We have, further, allowed that category systems may be revised compatibly with the demands of objective control. Finally, we have denied that alterations of theoretical viewpoint must destroy the very possibility of common meanings, of the reduction of laws, and of the accumulation of knowledge. We have, in effect, freed the concept of objectivity from its connections with fixity, and painted a picture of objective control as consistent with changing observations, languages, and beliefs.

We have, however, yet to consider a variety of arguments appealing to the history of science which have questioned the scientific commitment to objective controls, suggesting that the standard view sets forth an impossibly unrealistic, and hence irrelevant ideal for scientific practice. Such practice, studied in its concrete historical development, reveals the striking pervasiveness of personal and subjective factors in the growth of theory. The dominance of such factors exposes the objectivism of the standard view as an extravagant methodological myth. It remains for us to see whether criticisms of this sort can be adequately sustained by considerations drawn from the history of science.

Certain considerations occasionally adduced may be seen as irrelevant after the briefest reflection. The historical fact that scientists differ in personal characteristics and that they are, in particular, capable of wide variations of intellectual style does not in the least threaten objectivity as a feature of scientific culture. For such objectivity is primarily a matter of institutions which hold beliefs subject to public test by impartial criteria. To participate in the workings of such institutions is to acknowledge the critical relevance of impersonal considerations in the assessment of one's claims; it is to accept the responsibility of rational dialogue with others in the interests of truth, under the authority of observational credibility and logical cogency. Such participation is perfectly compatible with radical variation in qualities of intellect and personality. The scientific game imposes the constraints of descriptive accuracy, theoretical coherence, and logical discussion; it imposes no general limitations on passion, imagination, or flair.

Nor does scientific objectivity consist in a routine for theoretical discovery; if the scientist is not a disembodied and passionless intellect, neither is he a robot. He chooses and revises descriptive categories, guesses at connections and hidden factors, extrapolates selectively upon limited data, and invents new mechanisms, calculi, and theories, which may later be subjected to critical assessment. The Baconian hope of devising mechanical rules of scientific discovery has been largely abandoned, and Mill's methods of experimental inquiry have come to be properly seen as methods of testing hypotheses at best, not as methods of proof or discovery. Objectivity, in general, is a matter of test, control, and critique, and it characterizes the processing of ideas that have been independently and freely generated. Albert Einstein has put the main point as follows, in setting forth his "epistemological credo":

> The system of concepts is a creation of man together with the rules of syntax, which constitute the structure of the conceptual systems. Although the conceptual systems are logically entirely arbitrary, they are bound by the aim to permit the most nearly possible

certain (intuitive) and complete co-ordination with the totality of sense-experiences; secondly they aim at greatest possible sparsity of their logically independent elements (basic concepts and axioms), i.e., undefined concepts and underived propositions.[1]

We see here clearly the differentiation of theoretical creation from theoretical test. Creation is free; discipline enters in the evaluation of a theory's empirical adequacy, its logical coherence, and its relative simplicity. "All our thinking," says Einstein, "is of this nature of a free play with concepts; the justification for this play lies in the measure of survey over the experience of the senses which we are able to achieve with its aid."[2] Objectivity concerns the manner of justification; it requires only the responsible commitment to fair canons of control over one's theoretical claims. It does not also require what would indeed be contrary to the historical record, namely, that theories be manufactured according to some cookbook recipe.

It will be worthwhile to dwell a bit more on the distinction between the *generation* and the *evaluation* of theory, for the contrast will enable us to locate the most serious anti-objectivist arguments recently adduced on the basis of the history of science. This distinction has been strongly emphasized in the writings of Hans Reichenbach; he elaborates it in the first section of his book, *Experience and Prediction*. Concerned in this section to mark out the province of epistemology, Reichenbach begins by placing its descriptive task within the domain of sociology, addressed as it is, in the first instance, to knowledge as a sociological fact. Epistemology is nevertheless distinguishable from general sociology: epistemological description is concerned only with the internal structure and content of knowledge, whereas general sociology is interested also in the external relations into which knowledge enters, for example, the social and political setting within which a given science flourishes. To say that epistemological description is limited to the internal

[1] Albert Einstein, "Autobiographical Notes," tr. Paul Arthur Schilpp, in P. A. Schilpp, ed., *Albert Einstein: Philosopher-Scientist* (New York: Tudor Publishing Company, 1949), p. 13. Now published by the Open Court Publishing Company, La Salle, Illinois. By permission of the editor, translator, and publisher.

[2] *Ibid.*, p. 7.

structure of knowledge is, however, not sufficient to define its special focus. There is, in particular, a further need to separate it from psychology, since it might be supposed that "The internal structure of knowledge is the system of connections as it is followed in thinking." Reichenbach continues:

> From such a definition we might be tempted to infer that epistemology is the giving of a description of thinking processes; but that would be entirely erroneous. There is a great difference between the system of logical interconnections of thought and the actual way in which thinking processes are performed. The psychological operations of thinking are rather vague and fluctuating processes; they almost never keep to the ways prescribed by logic and may even skip whole groups of operations which would be needed for a complete exposition of the subject in question. That is valid for thinking in daily life, as well as for the mental procedure of a man of science, who is confronted by the task of finding logical interconnections between divergent ideas about newly observed facts; the scientific genius has never felt bound to the narrow steps and prescribed courses of logical reasoning.[3]

Reichenbach concludes that, whereas psychology deals with processes of thought as they occur in actuality,

> What epistemology intends is to construct thinking processes in a way in which they ought to occur if they are to be ranged in a consistent system; or to construct justifiable sets of operations which can be intercalated between the starting-point and the issue of thought-processes, replacing the real intermediate links. Epistemology thus considers a logical substitute rather than real processes.[4]

It is concerned, in short, with providing an idealization of actual thought, expressing in explicit and perspicuous manner its substance and logical form. Such an idealization Reichenbach calls, following Rudolf Carnap, a "rational reconstruction." It may be compared, he says,

> to the form in which thinking processes are communicated to other persons instead of the form in which they are subjectively per-

---

[3] Reprinted from *Experience and Prediction* by Hans Reichenbach by permission of the University of Chicago Press (Chicago: The University of Chicago Press, 1938), pp. 4–5.

[4] *Ibid.,* p. 5.

formed. The way, for instance, in which a mathematician publishes a new demonstration, or a physicist his logical reasoning in the foundation of a new theory, would almost correspond to our concept of rational reconstruction; and the well-known difference between the thinker's way of finding this theorem and his way of presenting it before a public may illustrate the difference in question. I shall introduce the terms *context of discovery* and *context of justification* to mark this distinction. Then we shall have to say that epistemology is only occupied in constructing the context of justification.[5]

Now a man's actual public presentation of his theory does not completely represent what Reichenbach intends by "rational reconstruction," for such presentation is itself typically lacking in full explicitness and rigor. What is offered here is only a comparison, but one which, as Reichenbach puts it, "may at least indicate the way in which we want to have thinking replaced by justifiable operations; and it may also show that the rational reconstruction of knowledge belongs to the descriptive task of epistemology. It is bound to factual knowledge in the same way that the exposition of a theory is bound to the actual thoughts of its author."[6]

The purpose of constructing a logically idealized version of thinking, however, is to facilitate its critical examination. Here the descriptive task of epistemology gives way to its critical task. The rational reconstruction of a given thought process may expose it to criticism in bringing unjustifiable chains of reasoning into view or in making clear that no justifiable connection between starting point and conclusion is available in context. The ultimate standpoint of epistemology is thus normative; it is concerned, in the last analysis, with assessing the legitimacy, that is, the material and logical persuasiveness, of thought. To this end, it addresses itself in the first instance not to the psychological background of the thought in question, nor to the process of its development, but rather to the ideal elaboration of its product, as a purportedly logical argument. The fact that such ideal elaboration is distinguishable from the actual course of thinking which is its crude base means that the very description of thought will be relevantly different as it is motivated by

[5]*Ibid.,* pp. 6–7.
[6]*Ibid.,* p. 7.

interest in the question of its justification rather than its origin or development.

Reichenbach presents his distinction, as we have seen, in the process of characterizing epistemology. But it is important to note that his characterization applies also to activities within science itself. There is, in general, I should argue, no sharp line between the concerns of science and the concerns of epistemology. Scientists themselves are continuously engaged in rational reconstruction, criticism, and evaluation of ideas within their respective domains of investigation. It is an oversimplification to present the man of science as purely a generator of new factual ideas, leaving to the epistemologist the tasks of idealization and normative criticism. Epistemology may, perhaps, be conceived as striving for a greater generality and explicitness in its formulations, but it builds upon modes of criticism and evaluation internal to the workings of science itself. To describe science *simply* as a process of criticism and evaluation would, to be sure, be a mistake; to describe it *simply* as a generative sequence of new theoretical ideas is, however, equally a mistake. For the critical appraisal of new ideas, in accord with accepted canons, influences their subsequent status within the community of science itself. The process of critical appraisal is, then, integral to science, its operative canons reflecting, not the course of theory generation, but rather the practice of theory assessment.

In general, therefore, if a given interpretation of objectivity, appealing to such canons, accords ill with the generation of scientific ideas, we have as yet no reason to condemn the attribution of such objectivity to science. For such attribution may well purport to characterize not the generation of theories, but their manner of justification, criticism, and control within the scientific realm. It may, to be sure, *idealize* such processes of evaluation in describing them, and it may be strongly linked with *approval* of such processes. At any rate, it is wrong to imply that, taken as purely descriptive, such attribution must either characterize actual processes of theoretical innovation, or else stand condemned as a species of science fiction or a piece of philosophical dogmatism. For it may in fact describe the institutionalized controls by which science itself evaluates theoretical novelty.

Nor is anything so far said intended to limit the normative judgment of the epistemologist, who remains free to evaluate scientific institutions in independent fashion, no matter how he describes them. He is, accordingly, free to uphold the ideal of objectivity even if he has a low opinion of the actual objectivity of scientific procedures. In his normative role, he exercises the prerogatives of the critic, and recognizes no duty to defer to scientific custom; his concern is with the legitimacy of thought rather than its actuality or effectiveness.

The main question that concerns us here, however, is whether objectivity can be upheld not simply as an abstract epistemological ideal but as an ideal which, regulating the characteristic activities of science, may enter into its very description. The distinction between theory genesis and theory evaluation, between the context of discovery and the context of justification, enables us to say with considerable plausibility that objectivity characterizes the evaluative or justificatory processes of science rather than the genesis of scientific ideas. Correlatively, we acquire with this distinction a strategy by which the historian's alleged counterexamples can be dealt with: they are, if possible, to be relegated to the domain of discovery and consigned thenceforth to psychology for further study. This has, indeed, been the dominant approach of the standard view.

Precisely this approach has been subjected to the most serious challenge in recent years by theorists basing their view upon the history of science. It has been suggested that the justificatory processes of science themselves fail of objectivity, that personal factors in actuality permeate not only the genesis of theory but also its evaluation, and that psychology is therefore crucially relevant to the explanation of both. The fundamental Reichenbachian distinction between the context of discovery and the context of justification has accordingly been rejected, together with his correlative distinction between epistemology and psychology, neither distinction being capable of saving objectivity as an actual feature of the processes of science. In place of the notion of Peirce that scientific convergence of belief is to be interpreted as a progressive revelation of reality, we are now to take such convergence as a product of rhetorical persuasion, psychological conversion, the natural elimination of unreconciled dissidents, and the retraining of the young by the victorious faction. Instead of reality's providing a constraint on scientific belief, reality is now to be seen as a projection of such belief, itself an outcome of non-rational influences. The central idealistic doctrine of the primacy of mind over external reality is thus resuscitated once again, this time in a scientific context.

It is worth remarking, at the outset, the striking self-contradictoriness that infects arguments for such a view when they themselves appeal, by way of rational justification, to the hard realities of the history of science. Yet, as in the case of the subjectivistic arguments from observation and from meaning earlier considered, we need to do more than simply point out such self-contradictoriness. We need to see how it is possible to avoid it ourselves, upon examination of the detailed considerations from which it flows. These considerations have appeared in one or another form, and variously combined, in several recent accounts. They are, however, best evaluated in the context of a single comprehensive presentation. Professor Thomas Kuhn's stimulating monograph, *The Structure of Scientific Revolutions,*[7] constitutes such a presentation, and has, moreover, been widely influential. Thus the ensuing discussion will focus on the points he raises, with an initial sketch of his treatment in order to fix our ideas.

Kuhn contrasts normal science with scientific revolutions. Normal science is based on certain past scientific achievements involving "law, theory, application, and instrumentation together" (p.

[7] All quotations from Kuhn in the text are reprinted from *The Structure of Scientific Revolutions* by Thomas S. Kuhn by permission of The University of Chicago Press (Chicago and London: The University of Chicago Press, 1962). Copyright under the International Copyright Union by The University of Chicago. All rights reserved.

Some important discussions relevant to the themes of the present lecture may be found in Michael Polanyi, *Personal Knowledge* (see note 14 above, p. 18); Polanyi's *The Tacit Dimension* (Garden City, New York: Doubleday & Company, Inc., 1966); and Leonard K. Nash, *The Nature of the Natural Sciences* (Boston: Little Brown and Company, 1963), esp. chaps. 9–12. For criticism of some of Polanyi's views, see Adolf Grünbaum, *Philosophical Problems of Space and Time* (New York: Alfred A. Knopf, 1963), pp. 377–386.

10). These achievements constitute "paradigms," that is, models of research embracing shared rules and standards of work defining coherent research traditions, and marking out special classes of problems yet to be solved.

> The success of a paradigm—whether Aristotle's analysis of motion, Ptolemy's computations of planetary position, Lavoisier's application of the balance, or Maxwell's mathematization of the electromagnetic field—is at the start largely a promise of success discoverable in selected and still incomplete examples. Normal science consists in the actualization of that promise, an actualization achieved by extending the knowledge of those facts that the paradigm displays as particularly revealing, by increasing the extent of the match between those facts and the paradigm's predictions and by further articulation of the paradigm itself [pp. 23–24].

The activity of normal science is puzzle-solving, a puzzle being both assured of a solution and restricted by rules, both features provided by the dominant paradigm (pp. 36–42). Normal science is cumulative but provides no novelties of fact or theory (p. 52). It may, however, lead to anomaly, which, when sufficiently pervasive, constitutes a crisis, "the persistent failure of the puzzles of normal science to come out as they should" (p. 68).

Confronted by crisis, scientists "do not renounce the paradigm that has led them into crisis" (p. 77). However, they may "begin to lose faith and then to consider alternatives" (p. 77). The activity of normal science begins to give way. For in this situation,

> scientists take a different attitude toward existing paradigms, and the nature of their research changes accordingly. The proliferation of competing articulations, the willingness to try anything, the expression of explicit discontent, the recourse to philosophy and to debate over fundamentals, all these are symptoms of a transition from normal to extraordinary research [p. 90].

A new paradigm emerges, sometimes foreshadowed by such extraordinary research, sometimes all at once "in the mind of a man deeply immersed in crisis" and in a manner that may remain forever inscrutable (p. 89). The transition to a new paradigm is a revolution.

Debates over paradigms, according to Kuhn, are characterized by an "incompleteness of logical contact" (p. 109) between the proponents of competing paradigms. Circular arguments abound, each paradigm being defended in terms of "the criteria that it dictates for itself" (p. 109), while disagreements among the competing schools as to what is a suitable problem and what is a solution, guarantee that "they will inevitably talk through each other when debating the relative merits of their respective paradigms" (p. 108). The choice between rival paradigms lies, in fact, beyond the capacities of normal science to resolve. Since it hinges upon considerations external to normal science, the issue in a paradigm debate is, indeed, revolutionary, involving a fundamental reconsideration and potential redefinition of normal science itself. Paradigm debates, moreover, reconstitute not only science, but nature as well (p. 109).

For the transition to a new paradigm occurs

> not by deliberation and interpretation, but by a relatively sudden and unstructured event like the gestalt switch. Scientists then often speak of the "scales falling from the eyes" or of the "lightning flash" that "inundates" a previously obscure puzzle, enabling its components to be seen in a new way that for the first time permits its solution [p. 121].

Since scientists see the world differently after a paradigm change, and since they have recourse to the world only through what they see and do, "we may want to say that after a revolution, scientists are responding to a different world" (p. 110). For example, we may well say that "after discovering oxygen, Lavoisier worked in a different world" (p. 117).

A new paradigm emerges first in a single mind, or in a few. "How are they able, what must they do, to convert the entire profession or the relevant professional subgroup to their way of seeing science and the world?" (p. 143). The transition from one to another paradigm is "a transition between incommensurables" and hence "cannot be made a step at a time. . . . Like the gestalt switch, it must occur all at once (though not necessarily in an instant) or not at all" (p. 149). Very often it never occurs. In any event, Kuhn suggests "that in these matters neither proof nor error is at issue. The transfer of allegiance from paradigm to paradigm is a conversion experience that cannot be forced" (p. 150). However, a general shift of paradigm may take place through a spread of conversions. "Conversions will occur a few at a time until, after the last holdouts have died, the whole profession will again be practicing under a single, but now a different, paradigm" (p. 151). And Kuhn cites Max Planck who, in his *Scientific Autobiography,* writes, "a new scientific truth does not triumph by convincing its opponents and making them see the light, but rather because its opponents eventually die, and a new generation grows up that is familiar with it" (p. 150).

What factors account for the change in viewpoint of the new generation? Some factors lie outside science altogether, some relate to personality, nationality, or reputation (p. 152). Other factors relate to claims that the new paradigm will solve the crisis problem or will prove more aesthetically appealing. But such claims cannot be sufficient; a decision is called for, and

> The man who embraces a new paradigm at an early stage must often do so in defiance of the evidence provided by problem-solving. He must, that is, have faith that the new paradigm will succeed with the many large problems that confront it, knowing only that the older paradigm has failed with a few. A decision of that kind can only be made on faith [p. 157].

The result of a scientific revolution must, however, be progress.

> Revolutions close with a total victory for one of the two opposing camps. Will that group ever say that the result of its victory has been something less than progress? That would be rather like admitting that they had been wrong and their opponents right. To them, at least, the outcome of revolution must be progress, and they are in an excellent position to make certain that future members of their community will see past history in the same way [p. 165].

Nevertheless, it is not wholly true that might makes right in science. The authority for choice of paradigm is vested in the scientific profession, which strives for general agreement, and which is concerned that the new paradigm should not lose previously acquired problem-solving ability and should seem, moreover, to solve new problems not otherwise soluble (pp. 166–68).

Having concluded my sketch of Professor Kuhn's ideas, I turn now to the task of evaluation. It is of interest to note, first, the striking way in which Kuhn's account applies psychological, political, and religious categories to the description of scientific change. The older references to logical system, observational evidence, theoretical simplicity, and experimental test have given way, in his account, to mention of the gestalt switch, conversion, faith, decision, and death. Moreover, such mention is introduced to characterize not only the initial generation of theory but also its subsequent spread and adoption by the scientific community. Indeed, such spread is described as consisting of a large series of individual conversion experiences—"the transfer of allegiance from paradigm to paradigm is a conversion experience," (p. 150)—producing "an increasing shift in the distribution of professional allegiances" (p. 157). The picture is rather like that of an epidemic.

Evaluative arguments over the merits of alternative paradigms are vastly minimized, such arguments being circular, and the essential factor consisting anyway not in deliberation or interpretation but rather in the gestalt switch. Nor does contrary evidence alone force rejection of a paradigm; such rejection occurs only if a suitable alternative paradigm is available. "No process yet disclosed by the historical study of scientific development," writes Kuhn, "at all resembles the methodological stereotype of falsification by direct comparison with nature" (p. 77). In fact, adoption of an alternative paradigm often requires actual defiance of the evidence and reliance on faith. We have here, it would seem, a radical rejection of the distinction between discovery and justification, in any sense at least that would preserve objective controls in the sphere of justification.

Some of the support for Professor Kuhn's view derives from arguments concerning observation and meaning such as those we have already considered and rejected in the last two lectures. What further considerations, what special emphases, are provided in Kuhn's treatment, leading him to minimize the importance of deliberation in paradigm change? To begin with, he certainly stresses the patterned alteration of thought and experience concomitant with a scientist's adoption of a new paradigm. Such an alteration, being "intuitive," "relatively sudden," and "unstructured" is more naturally assimilated to gestalt reorganizations of perception than to the "piecemeal" articulations associated with deliberation or interpretation (pp. 121–122).

This sort of consideration, however, supposing it to be factually true, seems strangely irrelevant to the critical issue. For this issue concerns not the psychology of the paradigm originator, but the public procedures of evaluation by which new paradigms are assessed. Let it be granted, for the sake of argument, that, as Kuhn declares, "No ordinary sense of the term 'interpretation' fits these flashes of intuition through which a new paradigm is born" (p. 122). It surely does not follow that the notion of interpretation also fails to apply to the processes by which the new paradigm is defended and criticized in the scientific community to which it is addressed.

If a new paradigm indeed effects something like a reorganization of perception, is such a reorganization typically accepted without argument as constituting a *self-evident* advance? Do scientists simply appeal to their own conversion experiences in *defending* a new paradigm? Even mathematics, the traditional home of intuition (though, oddly enough, largely ignored in Kuhn's treatment), does not plausibly yield affirmative answers to the foregoing questions. For mathematicians do not typically defend their intuitive convictions in print by simple appeal to their conversions or gestalt switches. Scientists, like mathematicians, may certainly be admitted to undergo experiences of sudden insight, but no more than mathematicians do they suppose that such experiences constitute justificatory reasons worthy of presentation to their unconvinced colleagues. The very existence of paradigm debates testifies, indeed, to their belief that independent supporting reasons are available to them, reasons which can sustain themselves in critical discussion of alternatives.

It might, of course, theoretically be the case that the scientists are systematically deluded about their own procedures, that their paradigm debates are always without substance and, at bottom, a mere interplay of rhetorical effects. However, one could certainly not show this by arguing simply from the psychology of perception, for the issue hinges on the content of the debates themselves. Nor would it be sufficient to adduce examples from the history of science of particular debates conducted at cross-purposes. For, in the first place, the existence of common evaluative criteria is compatible with borderline regions in which these criteria can yield no clear decisions. And, in the second place, even the objective availability of clear decisions is consistent with honest differences of judgment, not to mention plain misunderstandings.

What then can underlie Kuhn's assurance that paradigm change is not primarily a matter of deliberation and assessment but rather a matter of all-or-nothing conversions? One line of reasoning to this effect seems to rest on his contrast between normal science and what is external to normal science. "Paradigms," he writes, "are not corrigible by normal science at all . . . normal science ultimately leads only to the recognition of anomalies and to crises" (p. 121). Or, again, "the choice between competing paradigms regularly raises questions that cannot be resolved by the criteria of normal science. To the extent, as significant as it is incomplete, that two scientific schools disagree about what is a problem and what a solution, they will inevitably talk through each other when debating the relative merits of their respective paradigms" (p. 108). If significant debate based on normal science is thus ruled out on general grounds, it would seem plausible to have recourse to conversion as the ultimate determinant of paradigm change.

Such an argument seems to me, however, to be wholly specious. For normal science has initially been explained by Kuhn as just that sort of scientific research which is paradigm-bound. Is it any wonder, then, that the higher-level issue of paradigm choice outstrips the capability of normal science? Such a conclusion seems, indeed, a mere consequence of the initial construction of the phrase "normal science." To assume further, however, that deliberation and interpretation are *restricted* to normal science is to beg the very point at issue. For the question is precisely whether deliberation and interpretation are not also relevant to the extra-normal problem of paradigm choice. If paradigms are not corrigible by normal science, it does not follow that they are not corrigible at all. If scientific schools inevitably talk through each other when arguing from within their respective paradigms, it is not further inevitable that they do always argue from within their respective paradigms. There is, in fact, no good reason to conceive the resources of scientific schools as limited to those of normal science, and overwhelming plausibility to the supposition that scientists, no less than historians of science, for example, regularly transcend particular paradigms and address themselves to higher-order issues of paradigm comparison—such issues being, as is generally acknowledged, the substance of actual paradigm debates. The question then recurs: What independent reason can there be for supposing such debates to be mere persuasive displays without deliberative substance, in the face of contrary testimony by the participating scientists themselves?

Another answer is suggested by Kuhn's references to the so-called incommensurability of different paradigms. Competing paradigms are addressed to different problems; they embody different standards and even differing definitions of science (p. 147). They are based on different meanings and they operate in different worlds (pp. 148–149)—thus, "The proponents of competing paradigms are always at least slightly at cross-purposes" (p. 147). A fundamental idea suggested in such passages is that competing paradigms, like radically differing languages composed of wholly different elements, do not allow a point-by-point translation of content from one framework to the other. For apparently equivalent conceptual and experimental elements take on basically new roles under paradigmatic change. For example, as Kuhn remarks, "To make the transition to Einstein's universe, the whole conceptual web whose strands are space, time, matter, force, and so on, had to be shifted and laid down again on nature whole. . . . Communication across the revolutionary divide is inevitably partial" (p. 148). The conclusion seems to be that with step-by-step translation ruled out, conversion is the only alternative basis for paradigm change. As Kuhn writes,

> before they can hope to communicate fully, one group or the other must experience the conversion that we have been calling a paradigm shift. Just because it is a transition between incommensurables, the transition between competing paradigms cannot be made a step at a time, forced by logic and neutral experience [p. 149].

Paradigm debates cannot, then, be understood in terms of the categories of rational argument. They must fail to make logical or cognitive sense, owing to a fundamental failure of translation, and hence, of communication.

Now, there is an initial puzzle in the foregoing argument. If competing paradigms are indeed based in different worlds, and address themselves to different problems with the help of different standards, in what sense can they be said to be in competition?[8] How is it that there is any rivalry at all between them? To declare them in competition is, after all, to place them within *some* common framework, to view them within *some* shared perspective supplying, in principle at least, comparative and evaluative considerations applicable to both. It is, in fact, to consider them as oriented in different ways toward the same purposes, as making rival appeals from the standpoint of scientific goals taken to be overriding, and with respect to a common situation taken as a point of reference. But if this is indeed thought to be the case, a basis for reasonable comparison must in fact be presumed to exist, despite the incommensurability of the paradigms compared, and notwithstanding the fact that such a basis may fail, in particular cases, to decide clearly between paradigms in contention.

The main point to be stressed is that lack of commensurability, in the sense here considered, does not imply lack of comparability. Even works of art may be reasonably discussed, criticized, compared, and evaluated from various points of view; the incommensurability of these works does not establish that such critical discussion must consist of empty rhetoric alone. Comparison is not limited to an effort at what Kuhn calls "communication across the revolutionary divide"; it need not translate, step by step, one paradigm into some other, any more than art criticism need translate one work into another. Having appreciated the differing potentials of competing paradigms, the scientist, like the critic or indeed the historian, may step back and consider the respective bearings of

---

[8]As Dudley Shapere puts it, if paradigms "disagree as to what the facts are, and even as to the real problems to be faced and the standards which a successful theory must meet—then what are the two paradigms disagreeing about?" See the review of Kuhn by Shapere, "The Structure of Scientific Revolutions," *Philosophical Review*, LXXIII (1964), 391.

the paradigms with regard to issues he holds relevant. Such consideration is itself not formulated within, nor bound by, the paradigms which constitute its objects. It belongs rather to a second-order reflective and critical level of discourse. This is the level on which paradigm debates take place, and the incommensurability of their objects is no bar to their reasonableness or objectivity.

To be sure, such objectivity presupposes a certain sharing of standards at the second-order level if it is to characterize paradigm debates within the scientific *community,* and not simply the process of comparison within the mind of a single scientist. But is such second-order sharing of standards possible? Are not scientists of rival schools also isolated within incommensurable comparative and evaluative paradigms of the second order, among which transitions occur only as non-rational conversions? A single scientist may perhaps step back and rationally evaluate the respective merits of competing paradigms. But will not any two scientists of differing first-order persuasions inevitably step back to differing evaluative positions? Such a view seems to be suggested, at any rate, in certain passages of Kuhn's presentation:

> In learning a paradigm the scientist acquires theory, methods, and standards together, usually in an inextricable mixture. Therefore, when paradigms change, there are usually significant shifts in the criteria determining the legitimacy both of problems and of proposed solutions.
> That observation . . . provides our first explicit indication of why the choice between competing paradigms regularly raises questions that cannot be resolved by the criteria of normal science. . . . In the partially circular arguments that regularly result, each paradigm will be shown to satisfy more or less the criteria that it dictates for itself and to fall short of a few of those dictated by its opponent. There are other reasons, too, for the incompleteness of logical contact that consistently characterizes paradigm debates. For example, since no paradigm ever solves all the problems it defines and since no two paradigms leave all the same problems unsolved, paradigm debates always involve the question: Which problems is it more significant to have solved? Like the issue of competing standards, that question of values can be answered only in terms of criteria that lie outside of normal science altogether, and it is that recourse to external criteria that most obviously makes paradigm debates revolutionary [pp. 108–109].

We have earlier seen that the mere fact that paradigm choice lies beyond the scope of normal science does not show that such choice must lack deliberative substance. On the contrary, we have urged that the comparative evaluation of rival paradigms is quite plausibly conceived as a deliberative process occurring at a second level of discourse, that paradigm debates may indeed be considered to occur at this second level, regulated, to some degree at least, by shared standards appropriate to second-order discussion. The passage just quoted suggests, however, that such sharing of second-order standards is impossible. For to accept a paradigm is to accept not only theory and methods, but also governing standards or criteria which serve to justify the paradigm as against its rivals, in the eyes of its proponents. Paradigm differences are thus inevitably reflected upward, in criterial differences at the second level. It follows that each paradigm is, in effect, inevitably self-justifying, and that paradigm debates must fail of objectivity: again we appear driven back to non-rational conversions as the final characterization of paradigm shifts within the community of science.

This argument seems to me, however, to fail utterly, for it rests on a confusion. It fails to make the critical distinction between those standards or criteria which are internal to a paradigm, and those by which the paradigm is itself judged. The first paragraph of the passage just cited speaks of the paradigm as incorporating theory, methods, and standards, the latter representing criteria of legitimacy for scientific problems and proposed solutions. In Kuhn's discussion preceding the quoted passage, the attempt to explain gravity is offered as an example, such an attempt having been abandoned by eighteenth-century scientists but renewed successfully in the twentieth century by Einstein, thus returning science "to a set of canons and problems that are, in this particular respect, more like those of Newton's predecessors than of his successors" (p. 107). The sense, then, it would appear, in which standards are inextricably involved in a paradigm is just the sense in which the paradigm is understood as defining a range of acceptable problems and modes of solution within a particular scientific domain, providing scientists, as Kuhn remarks, "not only with a map but also with some of the directions essential for map-making" (p. 108).

So far, we are concerned with what I have called standards *internal* to a paradigm.

When we turn, however, to the second paragraph of the quoted passage, we are no longer considering standards of the same sort. The issue here is no longer the choice of problems or solutions within the scientific domain in question, but rather the choice of paradigms themselves. Our problem, in terms of Kuhn's metaphor, is not to construct a map of some region in accord with a given set of directions for map-making, but rather to compare and evaluate alternative sets of directions. Whatever the criteria by reference to which the latter task is carried out, it is clear that they are of a different order from, and independent of, the criteria embodied in any particular set of mapping rules. For no such set of rules implies how it is itself to be evaluated in comparison with alternative sets. *A fortiori,* no such set implies that it is itself to be ranked as superior to its alternatives.

The confusion of standards internal to a paradigm with those standards by which paradigms are themselves evaluated and compared is facilitated by an equivocal use of the word "criteria" in the quoted passage. Thus, the first paragraph speaks of "criteria determining the legitimacy both of problems and of proposed solutions," while the second paragraph, referring to paradigm debates, suggests that "each paradigm will be shown to satisfy more or less the criteria that it dictates for itself. . . ." The latter use of the word "criteria" clearly refers to standards by which paradigms are themselves judged, and it must be insisted that such standards are not dictated by, but are rather independent of the paradigms which constitute their range of application. As Kuhn rightly remarks toward the end of the quoted passage, the question of the relative significance of solved problems, like that of competing paradigmatic standards, "can be answered only in terms of criteria that lie outside of normal science." But then it is simply gratuitous to suppose that each paradigm "dictates" such second-order criteria—that, somehow inextricably involved in the paradigm, they serve always in turn to justify it as against its alternatives, thus inevitably precluding objectivity in the shared process of paradigm evaluation.

The argument that paradigms must be self-justifying thus collapses. It has failed completely to show that paradigm differences must be reflected upward in evaluative differences at the second level, that the sharing of second-order criteria by proponents of rival paradigms is impossible, and that the only recourse left to us is non-rational conversion. Sharing of evaluative criteria at the second-order level is, to be sure, not inevitable. But the considerations we have offered may be generalized: There is no necessity that second-order differences, where they occur, must themselves be reflected upward, no need to suppose that paradigm debates at any level must fail of objectivity and must consist ultimately of attempts at non-rational persuasion.

But still another reason is suggested by Kuhn for characterizing paradigm debates in terms of conversion rather than deliberation. This is the apparent resistance of paradigms to falsification. Were they construed as genuinely subject to test, they should be expected to fall with the discovery of counterinstances. In fact, however, argues Kuhn, nothing like this actually happens. Scientists "do not . . . treat anomalies as counterinstances, though in the vocabulary of philosophy of science that is what they are. . . . once it has achieved the status of paradigm, a scientific theory is declared invalid only if an alternate candidate is available to take its place" (p. 77). Rejection of the paradigm in crisis does not occur, then, until an alternative is available, while acceptance of the new paradigm defies the evidence of greater problem-solving ability of the old paradigm, and depends on faith (p. 157). "The competition between paradigms," says Kuhn, "is not the sort of battle that can be resolved by proofs" (p. 147).

Now acceptance in science is never a matter of *proof,* and the tentative acceptance of a relatively unsupported hypothesis is compatible with acknowledgment of controlling tests to which future experience will subject it. If this be faith, it is at least a faith which is consistent with objectivity of evaluation. Moreover, the notion of acceptance is ambiguous in a critical way: it need not always mean the taking of a proposition to be true. In the sciences, in particular, it may imply simply taking a hypothesis seriously enough to employ it as a leading principle of research and experimentation, without commitment as to its truth value. The crucial matter, from the point of view of a primary interest in objectivity, is not, in any event, the state of mind of "the man who embraces a new paradigm at an early stage" and who, according to Kuhn, "must often do so in defiance of the evidence provided by problem-solving" (p. 157). The crucial matter is rather the existence of shared scientific institutions of control by which paradigms, once adopted, are tested.

Insensitivity to available counterinstances is therefore indeed a serious issue. For such insensitivity conflicts with the control of hypotheses by reference to publicly ascertainable facts of observation and experimentation. To hold on to a theory stubbornly in the face of available counterinstances is to protect it from the test of experience. Such a policy would allow us to believe whatever we wish to believe, come what may.

Now, actually, this latter way of formulating the charge of insensitivity is ambiguous, and on one reading, clearly too strong. For if it means that a theory is typically retained along with statements acknowledged as formulating actual counterinstances to it, the charge is that scientists are willing to tolerate logical inconsistencies, a charge to which there is no plausibility whatever. The weaker version argues only that *purported* counterinstances to a favored theory are never accepted as *actual* counterinstances. They are explained away as due to error of one sort or another, their derivation is blocked by alteration of some subsidiary assumption, or they are simply rejected in the confidence that they are not genuine and that some alternative mode of dealing with them may perhaps some day be specified. This weaker version does not accuse scientists of tolerating inconsistency; it charges them rather with preserving consistency in a biased manner, that is, always in such a way as to protect their favored theories from overthrow. Now it may well be urged that system is central to the scientific quest, and that a certain favoring of a hard-won system, even at the cost of riding roughshod over a few purported facts, is still consistent with objective control. But a general and rigid insensitivity to putative counterinstances produced by observation and experiment would clearly be incompatible with such control.

Has Kuhn in fact shown such insensitivity to characterize scientific practice? He argues that the rejection of a paradigm in crisis always awaits a new paradigm. But the notion of rejection is as ambiguous as that of acceptance. It may mean rejection of a proposition as true, or loss of confidence in its truth. It may also mean rejection of a proposition as a guide to research, or as a practical tool. Now it is quite conceivable to reject a proposition as true but yet to continue to employ it in a limited way as a practical tool of research, experimentation, or technology. It is further conceivable, and consistent, to reject a proposition as true but to hold true a certain modification of it, incorporating a limitation that frees it from conflict with counterinstances. Such a course is perfectly compatible with objective control of belief and is often the most reasonable course to follow, especially in the absence of viable alternatives.

Relevant nuances in the notion of the rejection of scientific theories might well have been explored by Kuhn in comparison with parallel notions of the rejection of religious beliefs and political ideologies, especially in view of his extended comparisons with these systems in other respects. At any rate, Kuhn does seem to me implicitly to admit the critical point in allowing that, though scientists "do not renounce the paradigm that has led them into crisis," they "may begin to lose faith and then to consider alternatives" (p. 77), taking a "different attitude toward existing paradigms" (p. 90), expressing discontent, projecting "competing articulations" (p. 90), and even having "recourse to philosophy" (p. 90). Such response amounts to, or approximates, in fact, a rejection of the paradigm as true, but it does not therefore constitute a rejection in other senses of the word. The significance of this point is masked, I would suggest, by the global use of such notions as acceptance, rejection, and, last but not least, the term "paradigm" itself, which embraces not only theories but also modes of scientific practice.[9]

What compelling reasons have we then been offered for denying objectivity to the processes by which scientific theories are critically evaluated? We have, I believe, been offered none. In fact, it seems

[9]The vague and elastic use of the term "paradigm" is strongly criticized by Dudley Shapere in his review of Kuhn in *Philosophical Review,* pp. 383–394.

to me that Professor Kuhn himself concludes by reintroducing the very ideas he has been at pains to deny in the main tendency of his presentation. Despite his stress on incommensurability, for example, he writes that "the successful new theory must somewhere permit predictions that are different from those derived from its predecessor. That difference could not occur if the two were logically compatible" (p. 96). His purpose, in the context of this quotation, is to deny the cumulativeness of theories, but, in the course of this denial, he allows a predictive criterion as relevant to their comparative evaluation. He opposes received notions of falsification, but himself introduces the concepts of anomaly and crisis, which have a parallel function in his account. He downgrades the relevance of deliberation to paradigm change, but yet allows the importance of claims that the new paradigm will solve the problems that led the old one into crisis (p. 152). He criticizes the notion of cumulative science, yet insists that "though new paradigms seldom or never possess all the capabilities of their predecessors, they usually preserve a great deal of the most concrete parts of past achievement and they always permit additional concrete problem-solutions besides" (p. 168).

What is most important from the standpoint of the present lecture is that he seems to reinstate the very distinction between discovery and justification with which we started. For despite his strong emphasis on the conversion experience and the gestalt switch, he suggests several considerations relative to the critical evaluation of theories as actual elements of scientific functioning: the predictive criterion lately mentioned, the existence of anomaly and crisis, the preservation of previously acquired problem-solving abilities, and the promise "to resolve some outstanding and generally recognized problem that can be met in no other way" (p. 168). Such conditions of evaluation contradict the main thesis appealing to the history of science with which we have here been concerned, namely, that paradigm change in science is not generally subject to deliberation and critical assessment. They are also sufficient to show, I suggest, that objectivity is not a mere philosopher's dream, but an operative and controlling ideal of scientific practice.

5

Epistemology of Objectivity

We have been concerned, in the foregoing lectures, to delineate the threat of current subjectivism and to assess the soundness of its arguments. Our purpose has not been purely polemical, however. We have sought not an easy refutation of subjectivistic claims, but a refined analytical conception of the objective ideal against which such claims would be powerless. To this end, we have sketched an interpretation of various elements of objectivity in the context of our special discussions of observation, meaning, and change. In the present lecture, we address ourselves to the broad epistemological outlines of this interpretation in an effort to make clearer its distinctive character and its alternatives.

One such alternative has already been discussed and rejected, namely, the epistemology of phenomenal certainty, in the form represented by C. I. Lewis. The so-called certainty of the given, we argued, cannot protect its purported descriptions from mistake; the given can therefore not provide a fixed control over conceptualization. If we attempt to picture all our beliefs as somehow controlled by our reports of the given, we shall have to concede that these reports are themselves not rigidly constrained by what is given in fact, since they are themselves subject to error. It does no good, then, to suppose that they constitute points of direct and self-evident contact between our belief systems and reality—firm touchstones by which all our other beliefs are to be judged but which are themselves beyond criticism. Observation reports, in short, cannot be construed as isolated certainties. They must survive a continuous process of accommodation with our other beliefs, a process in the course of which they may themselves be overridden. The control they exercise lies not in an *infallibility* which is beyond their reach; it consists rather in an *independence* of other beliefs, an ability to clash with the rest in such a way as to force a systematic review threatening to all.

Such independence has, in fact, been the primary concern of our earlier discussions. We sought to show that the observational testing of a hypothesis is not necessarily a question-begging procedure, that observation need not be fatally contaminated by theoretical categories, that observation reports may, in consequence, perfectly well conflict with cherished hypotheses. But can such a conception of independence be sufficient for a theory of objective control over belief? Does it provide an adequate restriction of arbitrariness in the choice of hypotheses? Conflict provides at best, after all, a motivation for restoring consistency. However, if this is the only motivation I am bound to honor, I am free to choose at will among equally coherent bodies of belief at variance with one another; I need not prefer the consistent factual account to the consistent distortion nor, indeed, to the coherent fairy tale. Faced with a conflict between my observation reports and my theory, I may freely alter or discard the former or the latter or both, so long as I replace my initial inconsistent set of beliefs with one that is coherent. Clearly, this much freedom is too much freedom. Constraints beyond that of consistency must be acknowledged.

Yet, in denying the doctrine of certainty, have we not made it impossible to do just that? If all our beliefs are infected with the possibility of error, if none of our descriptions is guaranteed to be true, none can provide us with an absolutely reliable link to reality. None can serve, through an immediately transparent correspondence with fact, as an additional, referential constraint upon our choices of belief. Our beliefs float free of fact, and the best we can do is to ensure consistency among them. The dilemma is severe and uncomfortable: swallow the myth of certainty or concede that we cannot tell fact from fancy.

This dilemma lies at the root of much controversy among scientifically minded philosophers in recent decades. A review of certain elements of the controversy will enrich our grasp of the problem and help to elucidate the approach of these lectures. We take as the primary object of such review the debate within the Vienna Circle in the nineteen-thirties concerning the status of so-called protocol sentences in science. Two chief protagonists in this debate were Otto Neurath and Moritz Schlick, the former rejecting the doctrine of certainty and insisting "that science keeps within the domain of propositions, that propositions are its starting point and terminus,"[1] and the latter urging rather that science is "a means of finding one's way among the facts," its confirmation-statements constituting "absolutely fixed points of contact" between "knowledge and reality."[2]

Let us turn first to Neurath who, in his anti-metaphysical zeal, proposes not only that science be purged of phenomenalism and unified through expression in physicalistic language, but also that scientific operations be understood as wholly confined to the realm of statements:

> It is always science as a system of statements which is at issue. *Statements are compared with statements,* not with "experiences," "the world," or anything else. All these meaningless *duplications* belong to a more or less refined metaphysics and are, for that reason,

[1] Otto Neurath, "Sociology and Physicalism," tr. Morton Magnus and Ralph Raico. Reprinted with permission of The Free Press from *Logical Positivism* by A. J. Ayer, ed., p. 285. Copyright © 1959 by The Free Press, A Corporation. Originally appeared as "Soziologie im Physikalismus," *Erkenntnis,* II (1931/2). Page references to this article in the text will be preceded by "SP."

[2] Moritz Schlick, "The Foundation of Knowledge," tr. David Rynin. Reprinted with permission of The Free Press from *Logical Positivism* by A. J. Ayer, ed., p. 226. Copyright © 1959 by The Free Press, A Corporation. Originally appeared as "Über das Fundament der Erkenntnis," *Erkenntnis,* IV (1934). Page references to Schlick in the text refer to this article.

to be rejected. Each new statement is compared with the totality of existing statements previously coordinated. To say that a statement is correct, therefore, means that it can be incorporated in this totality. What cannot be incorporated is rejected as incorrect. The alternative to rejection of the new statement is, in general, one accepted only with great reluctance: the whole previous system of statements can be modified up to the point where it becomes possible to incorporate the new statement. . . . The definition of "correct" and "incorrect" proposed here departs from that customary among the "Vienna Circle," which appeals to "meaning" and "verification." In our presentation we confine ourselves always to the sphere of linguistic thought [SP, 291].

Against the notion of a primitive and incorrigible set of so-called protocol statements as the basis of science, Neurath is adamant. "There is no way of taking conclusively established pure protocol sentences as the starting point of the sciences," he writes.[3] Aside from tautologies, the protocol as well as the non-protocol sentences of unified science share the same physicalistic form and are subject to the same treatment. The protocol statements are distinguished by the fact that "in them, a personal noun always occurs several times in a specific association with other terms. A complete protocol sentence might, for instance, read: 'Otto's protocol at 3:17 o'clock: [At 3:16 o'clock Otto said to himself: (at 3:15 o'clock there was a table in the room perceived by Otto)]'" (PS, 202). However, the main point to be stressed is not that protocol sentences are distinct but rather that *"Every law and every physicalistic sentence of unified-science or of one of its sub-sciences is subject to . . . change. And the same holds for protocol sentences"* (PS, 203).

The motivation for change is the wish to maintain consistency, for "In unified science we try to construct a non-contradictory system of protocol sentences and non-protocol sentences (including laws)" (PS, 203). Thus it is that a new sentence in conflict with the accepted system may dislodge a systematic sentence or may

[3]Otto Neurath, "Protocol Sentences," tr. Frederic Schick. Reprinted with permission of The Free Press from *Logical Positivism* by A. J. Ayer, ed., p. 201. Copyright © 1959 by The Free Press, A Corporation. Originally appeared as "Protokollsätze," *Erkenntnis*, III (1932/3). Page references to this article in the text will be preceded by "PS."

itself be rejected, and "The fate of being discarded may befall even a protocol sentence" (PS, 203).

The notion that protocol sentences are primitive and beyond criticism because they are free of interpretation must be abandoned, for "The above formulation of a complete protocol sentence shows that, insofar as personal nouns occur in a protocol, interpretation must *always* already have taken place" (PS, 205). Furthermore, there is, within the innermost brackets, an inescapable reference to some person's "act of perception" (PS, 205). The conclusion is that no sentence of science is to be regarded as more primitive than any other:

> All are of equal primitiveness. Personal nouns, words denoting perceptions, and other words of little primitiveness occur in all factual sentences, or, at least, in the hypotheses from which they derive. All of which means that *there are neither primitive protocol sentences nor sentences which are not subject to verification* [PS, 205].

Further, since *"every* language *as such,* is inter-subjective" (PS, 205), it is meaningless to talk of private languages, or to regard protocol languages as initially disparate, requiring ultimately to be brought together in some special manner. On the contrary, "The protocol languages of the Crusoe of yesterday and of the Crusoe of today are as close and as far apart from one another as are the protocol languages of Crusoe and of Friday" (PS, 206).

> Basically, it makes no difference at all whether Kalon works with Kalon's or with Neurath's protocols, or whether Neurath occupies himself with Neurath's or with Kalon's protocols. In order to make this quite clear, we could conceive of a sorting-machine into which protocol sentences are thrown. The laws and other factual sentences (including protocol sentences) serving to mesh the machine's gears sort the protocol sentences which are thrown into the machine and cause a bell to ring if a contradiction ensues. At this point one must either replace the protocol sentence whose introduction into the machine has led to the contradiction by some other protocol sentence, or rebuild the entire machine. *Who* rebuilds the machine, or *whose* protocol sentences are thrown into the machine is of no consequence whatsoever. Anyone may test his own protocol sentences as well as those of others [PS, 207].

Neurath stresses the place of prediction in science. He argues against phenomenal language that it "does not even seem to be usable for 'prediction'—the essence of science . . . " (SP, 290), and urges in favor of physicalism that it enables us to "achieve successful predictions" (SP, 286). He hopes that the fruitfulness of social behaviorism will be shown by the "successful predictions" it will yield (SP, 317), and looks forward to the day when a physicalistic sociology will "formulate valid predictions on a large scale" (SP, 317). Yet, true to his self-imposed restriction to the realm of statements alone, he does not construe the success of a prediction as consisting in its agreement with fact. Rather, he declares: "A prediction is a statement which it is assumed will agree with a future statement" (SP, 317).

Despite his refusal, however, to contrast the "thinking personality" with "experience" (SP, 290), to compare statements with " 'experiences,' 'the world,' or anything else" (SP, 291), and to ask such " 'dangerous' questions . . . as how 'observation' and 'statement' are connected; or, further, how 'sense data' and 'mind,' the 'external world' and the 'internal world' are connected,"[4] he slips into what he ought surely to have regarded, in a more careful moment, as dangerous metaphysics:

> Ignoring all meaningless statements, the unified science proper to a given historical period proceeds from proposition to proposition, blending them into a self-consistent system which is an instrument for successful prediction, and, consequently, for life [SP, 286].

To speak rashly in this way of the relation between science and life is clearly to leave the pure realm of statements and to admit, after all, that science cannot be adequately characterized in terms of consistency alone, that its very point, indeed, is to refer to what lies beyond itself.

Surely, not all self-consistent systems are "instruments for life," in the intended sense. The supposition that unified science issues in such practically useful instruments goes beyond the range of consistency in a manner that is not satisfactorily explained by

---

[4]Otto Neurath, *Foundations of the Social Sciences* (Chicago: The University of Chicago Press, 1944), p. 5.

Neurath's general account. He implies, of course, that practical usefulness accrues to science in virtue of its yielding successful predictions. This explanation is hardly adequate, however, for Neurath understands the success of a prediction to consist simply in its agreement with a later statement; on this criterion all predictions succeed which are followed by reiterations of themselves or by other statements coherent with them.

In the first of the passages by Neurath earlier quoted, he speaks of comparing each new statement with "the totality of existing statements previously coordinated" (SP, 291), to determine whether or not the statement can be incorporated in the totality. Perhaps the idea is that there is one presumably coherent totality which is to be singled out as a standard on each occasion of comparison, namely, that totality last ratified by acceptance and still in force on that occasion. The factor of *acceptance* may thus be thought to constitute a relevant selective consideration beyond consistency—a consideration that, moreover, escapes the dangers of metaphysics by avoiding appeal to a reality to which statements refer.

Now the *acceptance* of a statement is indeed relevantly independent of its *reference,* but acceptance also fails to differentiate between beliefs that are critically accepted on the basis of factual evidence and those that are not. The method of comparison recommended in the passage under consideration thus applies as well to entrenched myths and indoctrinated distortions as to scientific systems. What is of crucial significance, however, is that this method provides no incentive to *revise* the accepted totality of beliefs. For the assumed coherence of this totality can always be preserved by rejecting *all* new conflicting sentences. Neurath concedes, in fact, that the alternative to such rejection, consisting in revision of the accepted totality, is adopted "only with great reluctance" (SP, 291). The mystery, on his account, is why it should ever be adopted at all. Any coherent totality is, so far as his method is concerned, capable of being established as forever safe from revision, and thereby warranted as correct, to boot.

It is, moreover, pertinent to question the assumed interpretation of acceptance: acceptance by whom? The assumption that acceptance singles out one presumably coherent totality on each occasion of comparison is perhaps plausible if we consider just one individual. It is groundless if we take into account the acceptances of the whole "inter-subjective" community in line with Neurath's general attitude. For he deplores the "emphasis on the 'I' familiar to us from idealistic philosophy" (PS, 206), and considers it meaningless to talk of personal protocol languages. "One can," he writes, "distinguish an *Otto-protocol* from a *Karl-protocol,* but not a protocol of one's own from a protocol of others" (PS, 206). A general rather than an individual appeal to the factor of acceptance, however, yields a multiplicity of conflicting totalities of belief: Which of these is to serve as a standard?

There are passages, indeed, in which Neurath seems to be making no appeal to anything even approximating acceptance. He does not speak, in these passages, of *"the* totality of existing statements previously coordinated," but acknowledges rather a plurality of mutually conflicting totalities simply as abstract choices open to the investigator. To be consistent, the investigator may not choose more than one of these, but there is no further constraint on his choice beyond convenience. Thus, Neurath writes:

> A social scientist who, after careful analysis, rejects certain reports and hypotheses, reaches a state, finally, in which he has to face comprehensive sets of statements which compete with other comprehensive sets of statements. All these sets may be composed of statements which seem to him plausible and acceptable. There is no place for an empiricist question: Which is the "true" set? but only whether the social scientist has sufficient time and energy to try more than one set or to decide that he, in regard to his lack of time and energy—and this is the important point—should work with one of these comprehensive sets only.[5]

We find here, to be sure, a passing reference to plausibility and acceptability, but it is wholly unexplained, and can, moreover, have no point unless we move outside the "sphere of linguistic thought" (SP, 291) in a manner for which Neurath has altogether failed to prepare us. As to the choice among incompatible systems, any one is as good as any other; within the limits of time and energy,

[5]*Foundations of the Social Sciences,* p. 13.

the decision between them is arbitrary. The machine analogy earlier quoted does indeed, as Neurath says, make the point "quite clear." The machine detects contradictions but, aside from a general restriction to physicalistic language which may be assumed, no principle of selection is supplied for determining its input. Protocol sentences, distinguished solely by their form, may be chosen arbitrarily for insertion. Nor is there any restriction on the structure of the machine beyond its requiring the inclusion of at least some laws, presumably also distinguished by their form alone. So long as no contradiction has been detected among its virtually arbitrary elements, moreover, the machine is to be taken as the very embodiment and standard of correctness. The picture is one of unrelieved coherence free of any taint of fact. Since any consistent statement or system whatever can be accommodated by some such machine, any such statement or system can be fastened upon, held to be correct, and thenceforth protected forever from revision. The dogmatism of certainty has given way to the dogmatism of coherence. We have here not a picture of science but a desperate philosophical caricature.

What impels Neurath to construct this caricature? To appreciate his philosophical motivation is to gain a deeper understanding of the basic dilemma we face between coherence and certainty. He is, as we have seen, opposed to the idea that protocol sentences are above criticism because totally free of interpretation, serving simply to register the raw facts as given. On the contrary, he upholds the view that all statements in science are subject to change, insisting that observation reports may themselves be discarded under pressure of conflict with other scientific statements. Accordingly he emphasizes the "unified" nature of science, that is to say, the fact that no statement is an island—that each can survive only within a systematically harmonious community of statements. In place of the doctrine that selected statements provide an infallible contact with reality and are thus privileged to exercise unilateral control over the rest, Neurath urges a fluid and egalitarian conception: control is provisional, mutual, and diffused throughout the community of statements, resting in no case upon infallible access to fact.

Indeed, the very notion of such access seems to require the supposition that statement and reality might, through direct comparison, be determined to correspond with each other. But such a supposition is meaningless from Neurath's point of view. One can certainly compare statements with statements, but to imagine that statements can be literally compared with reality or with facts is to fall prey to an obfuscating metaphysics.

Now Neurath's remarks on this theme may appear, at first blush, to be simply a dogmatic denial of the obvious. His underlying thought may perhaps be interpreted more plausibly as a rejection of the philosophical tendency to read linguistic features into reality. The structure of language is not, after all, to be taken naively as a clue to the structure of reality. The correspondence suggested, for example, between atomic statements and atomic facts, and between molecular statements and molecular facts, is supported by nothing more than an anterior, and quite gratuitous hypostatization of objects to which certain elements of a language may be said to be directed. Facts, in general, understood as peculiar extra-linguistic entities precisely parallel to true statements, belong, in Neurath's scheme, to the class of "meaningless duplications . . . to be rejected" (SP, 291).[6]

Not only are such duplications superfluous; they mislead us into supposing that, in locating them independently and finding them to share the same structure with certain statements, we have a genuine method of justifying the acceptance of these statements. But facts, as entities distinct from the true statements to which they are presumed to correspond, have no careers of their own capable of sustaining such a method. These ghostly copies of true statements cannot be independently specified, confronted, or analyzed; their reality is no easier to determine than the truth of their respective parent sentences. Faced with the problem whether to *judge* a given

---

[6]On the parallelism of language and reality, see Ludwig Wittgenstein, *Tractatus Logico-Philosophicus* (London: Routledge & Kegan Paul, Ltd., 1922). For further discussion, see Edna Daitz, "The Picture Theory of Meaning," *Mind* (1953), reprinted in A. G. N. Flew, ed., *Essays in Conceptual Analysis* (London: Macmillan & Co., Ltd., 1960); John Passmore, *A Hundred Years of Philosophy* (London: Gerald Duckworth & Co. Ltd., 1957); and Nelson Goodman, "The Way the World Is," *Review of Metaphysics,* XIV (1960), 48–56.

sentence *as* true, it therefore does us no good to be told simply to ascertain whether there exists a structurally corresponding fact. If I am undecided about the truth of the sentence "The car is in the garage," I am equally undecided as to whether or not it is a fact that the car is in the garage: there are not two issues here, but one. Nor do I see how to go about resolving the latter indecision in a way that differs from my attempt to resolve the former. Appeal to the facts, taken strictly, thus turns out question-begging as a general method for ascertaining truth. For it requires, in effect, that the truth be determined as a condition of its own ascertainment.

The import of this line of reasoning may be illustrated strikingly by a consideration of prediction. The prevalent view is that in science, at any rate, a set of beliefs is put to the test by deriving therefrom a prediction that can be checked observationally against actual experience. When such a prediction is borne out by experience, the set of beliefs in question has passed a critical test; when the prediction is violated by experience, the test has been failed and the set must thereupon be revised so as to eliminate the prediction in question. The question that needs to be faced, however, is the question of how we can tell whether or not a prediction has been borne out or violated by experience. It must be stressed to begin with that the relations of logical consistency and contradiction hold only between certain statements and others, and *not* between statements and experiences. It may well be granted, therefore, that if a system S yields a prediction P and if we independently require S to be logically consistent with *Not-P,* we shall have to revise S. So far the issue concerns only the consistency relations among statements. But what, it may be asked, is our initial basis for setting logical consistency with *Not-P* as a constraint upon S? What can lead us to adopt *Not-P* in the first place?

To appeal to the logical consistency of *Not-P* with experience is nonsense. To say that we accept *Not-P* if it in turn yields predictions that are borne out by experience is to take the fatal first step in an infinite regress. To suggest that *Not-P* be judged true if and only if the corresponding fact represented by *Not-P* is real begs the question, as we have seen. To suppose, finally, that we have an infallible intuition of the truth of *Not-P* as a description of reality, that somehow its truth is immediately and indubitably evident to us upon intellectual inspection, is to revert to the myth of certainty. The conclusion to which we thus appear driven is that the whole idea of checking beliefs against experience is misguided. We do not go outside the realm of statements at all. What figures in the control of our system of beliefs is not experience, but purported statements of experience; not observation, but observation reports. Such is Neurath's conclusion, as we have already seen—a conclusion that, however well motivated, must surely be judged unacceptable as an account of science.

Convinced of the unacceptability of Neurath's account, Schlick insists that there must be an "unshakeable point of contact between knowledge and reality" (p. 226). To give up "the good old expression 'agreement with reality' " (p. 215), and to espouse instead a coherence theory such as that propounded by Neurath yields intolerable consequences:

> If one is to take coherence seriously as a general criterion of truth, then one must consider arbitrary fairy stories to be as true as a historical report, or as statements in a textbook of chemistry, provided the story is constructed in such a way that no contradiction ever arises. I can depict by help of fantasy a grotesque world full of bizarre adventures: the coherence philosopher must believe in the truth of my account provided only I take care of the mutual compatibility of my statements, and also take the precaution of avoiding any collision with the usual description of the world, by placing the scene of my story on a distant star, where no observation is possible. Indeed, strictly speaking, I don't even require this precaution; I can just as well demand that the others have to adapt themselves to my description; and not the other way round. They cannot then object that, say, this happening runs counter to the observations, for according to the coherence theory there is no question of observations, but only of the compatibility of statements.
>
> Since no one dreams of holding the statements of a story book true and those of a text of physics false, the coherence view fails utterly. Something more, that is, must be added to coherence, namely, a principle in terms of which the compatibility is to be established, and this would alone then be the actual criterion [pp. 215–216].

Since, in the case of conflict within a given set of statements, the coherence theory allows us to eliminate such conflict in various ways, "on one occasion selecting certain statements and abandoning or altering them and on another occasion doing the same with the other statements that contradict the first," the theory fails to provide an unambiguous criterion, yielding "any number of consistent systems of statements which are incompatible with one another" (p. 216). Schlick concludes that "The only way to avoid this absurdity is not to allow any statements whatever to be abandoned or altered, but rather to specify those that are to be maintained, to which the remainder have to be accommodated" (p. 216).

One might suppose, on the basis of such a conclusion, that Schlick would proceed to a defense of the certainty of protocol statements. Not so, however. He grants that such statements, as exemplified by familiar recorded accounts of scientific observation, and associated with "empirical facts upon which the edifice of science is subsequently built," are indeed subject to error and revision. "They are anything but incontrovertible, and one can use them in the construction of the system of science only so long as they are supported by, or at least not contradicted by, other hypotheses" (pp. 212–213). Even our own previously enunciated protocol statements may be withdrawn. "We grant," writes Schlick,

> that our mind at the moment the judgment was made may have been wholly confused, and that an experience which we now say we had two minutes ago may upon later examination be found to have been an hallucination, or even one that never took place at all.
>   Thus it is clear that on this view of protocol statements they do not provide one who is in search of a firm basis of knowledge with anything of the sort. On the contrary, the actual result is that one ends by abandoning the original distinction between protocol and other statements as meaningless [p. 213].

Schlick thus agrees with Neurath in denying a privileged role to protocol statements. Like Neurath, he insists that they "have in principle exactly the same character as all the other statements of science: they are hypotheses, nothing but hypotheses" (p. 212). Where, then, is the fixed point of contact between knowledge and reality? Schlick's view is that it is to be located in a special class of statements that are not themselves within science but are nevertheless essential to its function and, in particular, to its confirmation. His special term for these statements is *Konstatierungen,* though he sometimes calls them "observation statements"; I shall here refer to them uniformly as "confirmation statements."[7]

A confirmation statement is a momentary description of what is simultaneously perceived or experienced. It provides an *occasion* for the production of a protocol statement proper, which is preserved in writing or in memory; it must, however, be sharply distinguished from the protocol statement to which it may give rise. For this protocol statement can no longer describe what is simultaneous with itself; the critical experience has lapsed during the time taken to fix it in writing or memory. The protocol statement, moreover, unlike the confirmation statement, does not die as soon as it is born; its own life extends far beyond the initial point nearest the experience in question. Though it has, to be sure, the advantage of providing an enduring account, the protocol statement is, thus, never more than a hypothesis, subject to interpretation and revision. "For, when we have such a statement before us, it is a mere assumption that it is true, that it agrees with the observation statements [i.e., the confirmation statements] that give rise to it" (pp. 220–221).

Confirmation statements may serve to stimulate the development of genuine scientific hypotheses, but they are too elusive to be construed as the ultimate and certain *basis* of knowledge. Their contribution consists rather in providing an absolute and indubitable culmination to the process of testing hypotheses. When a predicted experience occurs, and we simultaneously pronounce it to have occurred, we derive "thereby a feeling of *fulfilment,* a quite characteristic satisfaction: we are *satisfied*" (p. 222). Confirmation

---

[7]There are problems in choosing a suitable translation; see David Rynin's note on these problems in Ayer, ed., *Logical Positivism,* p. 221. I choose "confirmation statements" to emphasize that statements are thus denoted, in preference to Rynin's "confirmations," though I believe the latter choice follows Schlick's own usage more closely.

statements perform their characteristic function when we obtain such satisfaction.

> And it is obtained in the very moment in which the confirmation takes place, in which the observation statement [i.e., confirmation statement] is made. This is of the utmost importance. For thus the function of the statements about the immediately experienced itself lies in the immediate present. Indeed we saw that they have so to speak no duration, that the moment they are gone one has at one's disposal in their place inscriptions, or memory traces, that can play only the role of hypotheses and thereby lack ultimate certainty. One cannot build any logically tenable structure upon the confirmations, for they are gone the moment one begins to construct. If they stand at the beginning of the process of cognition they are logically of no use. Quite otherwise however if they stand at the end; they bring verification (or also falsification) to completion, and in the moment of their occurrence they have already fulfilled their duty. Logically nothing more depends on them, no conclusions are drawn from them. They constitute an absolute end [p. 222].

In bringing a cycle of testing to an absolute close, a confirmation statement helps to steer the further course of scientific investigation: a falsified hypothesis is rejected and the search for an adequate replacement ensues; a verified hypothesis is upheld and "the formulation of more general hypotheses is sought, the guessing and search for universal laws goes on" (p. 222). The cognitive culmination represented by confirmation statements had, originally, according to Schlick, a purely practical import: it indicated the reliability of underlying hypotheses as to the nature of man's environment, and thus aided man's adjustment to this environment. In science, the joy of confirmation is no longer tied to the "purposes of life" (p. 222), but is pursued for its own sake:

> And it is this that the observation statements [confirmation statements] bring about. In them science as it were achieves its goal: it is for their sake that it exists. . . . That a new task begins with the pleasure in which they culminate, and with the hypotheses that they leave behind does not concern them. Science does not rest upon them but leads to them, and they indicate that it has led correctly. They are really the absolute fixed points; it gives us joy to reach them, even if we cannot stand upon them [p. 223].

What is it, however, that enables confirmation statements to constitute "absolute fixed points"? In what does their special claim to certainty consist? Schlick conceives these statements as always containing demonstrative terms. His examples are, "Here yellow borders on blue," "Here two black points coincide," "Here now pain." The constituent demonstratives function as gestures. "In order therefore to understand the meaning of such an observation statement [confirmation statement] one must simultaneously execute the gesture, one must somehow point to reality" (p. 225). Thus, he argues, one can understand a confirmation statement "only by, and when, comparing it with the facts, thus carrying out that process which is necessary for the verification of all synthetic statements" (p. 225). For to comprehend its meaning is simultaneously to apprehend the reality indicated by its demonstrative terms.

> While in the case of all other synthetic statements determining the meaning is separate from, distinguishable from, determining the truth, in the case of observation statements [confirmation statements] they coincide. . . . the occasion of understanding them is at the same time that of verifying them: I grasp their meaning at the same time as I grasp their truth. In the case of a confirmation it makes as little sense to ask whether I might be deceived regarding its truth as in the case of a tautology. Both are absolutely valid. However, while the analytic, tautological, statement is empty of content, the observation statement [confirmation statement] supplies us with the satisfaction of genuine knowledge of reality [p. 225].

The distinctiveness of confirmation statements lies, then, in their immediacy, that is, their capacity to point to a simultaneous experience, in the manner of a gesture. To such immediacy they "owe their value and disvalue; the value of absolute validity, and the disvalue of uselessness as an abiding foundation" (p. 225). It is of the first importance, for Schlick's view, to recognize the distinctiveness of confirmation statements and, in particular, to separate them from protocol statements, for this separation is the key to the problem as he sees it. "Here now blue" is thus not to be confused with the protocol statement of Neurath's type: "M.S. perceived blue on the nth of April 1934 at such and such a time and such and such a place."

The latter is an uncertain hypothesis, but it is distinct from the former: it must mention a perception and identify an observer. On the other hand, one cannot write down a confirmation statement without altering the meaning of its demonstratives, nor can one formulate an equivalent without demonstratives, for one then "unavoidably substitutes . . . a protocol statement which as such has a wholly different nature" (p. 226).

In sum, if we consider simply the body of scientific statements, they are all hypotheses, all uncertain. To take into account also the relation of this body of statements to reality requires, however, that we acknowledge the special role of confirmation statements as well. An understanding of these statements enables us to see science as "that which it really is, namely, a means of finding one's way among the facts; of arriving at the joy of confirmation, the feeling of finality" (p. 226). These statements do not "lie at the base of science; but like a flame, cognition, as it were, licks out to them, reaching each but for a moment and then at once consuming it. And newly fed and strengthened, it flames onward to the next" (p. 227).

We have already expressed our own agreement with the critical side of Schlick's doctrine, namely, his rejection of a coherence theory such as Neurath's. We, too, have stressed the importance of acknowledging constraints upon our belief beyond those imposed by consistency alone. We can therefore sympathize with Schlick's effort to propose an account of scientific systems that relates them to the reality to which they purport to refer. And it must indeed be admitted that the spirit of his general view of science is in closer accord than Neurath's with our familiar conceptions as well as with the understandings of scientists themselves as to the purport of their own activities.

Yet Schlick's positive theory suffers from a variety of fundamental difficulties that render it altogether unacceptable. Let us consider, first of all, whether his doctrine of confirmation statements is capable of meeting the problem as he has diagnosed it. He seeks, after all, a principle beyond coherence "in terms of which the compatibility is to be established," insisting that the only way to avoid the difficulties of the coherence theory is to avoid allowing "any statements whatever to be abandoned or altered, but rather to specify those that are to be maintained, to which the remainder have to be accommodated" (p. 216).

But if this is indeed the only way to avoid the difficulties of the coherence view, then it must be doubted that Schlick's positive doctrine in fact succeeds in avoiding them. For the coherence view purports to be a theory of science—of "science as a system of statements," as Neurath puts it. Any attempt to restrict the arbitrariness of coherence along the lines of Schlick's diagnosis must specify fixed points to which the statements of science are to be adjusted. It must, that is, specify a fixity to which *science* is responsive, by which *scientific* spontaneity is contained. In particular, it must not permit every scientific statement whatever to be subject to revision but must, on the contrary, place definite limits upon statement revision within science.

It is just here that Schlick's doctrine fails. For he identifies as "absolute fixed points" only *confirmation statements,* which fall outside science, and he insists, moreover, that these statements provide no barrier whatever to the revision of scientific statements proper. In particular, Schlick stresses that protocol statements, which are the closest counterparts of confirmation statements within science, "have in principle exactly the same character as all the other statements of science: they are hypotheses, nothing but hypotheses . . . one can use them in the construction of the system of science only so long as they are supported by, or at least not contradicted by, other hypotheses" (pp. 212–213). We have here, it seems, a clear admission that, within the realm of science, coherence continues to rule, despite the certainty attributed to confirmation statements. The latter have in effect been so sharply sundered from the body of science that they can yield it no advantage derived from their own presumed fixity. If reality alone provides no fixed control over scientific systems, the postulation of intermediate confirmation statements thus accomplishes nothing in the way of achieving such control.

Nor is it easy to make Schlick's general account of the scientific role of such statements intelligible. They are described as having an essential role in scientific functioning—in particular, in the testing and verification of hypotheses. They are not to be thought of as constituting a logical basis or origin of science. "If they stand at the beginning of the process of cognition they are logically of no use" (p. 222). Rather, "they bring verification (or also falsification) to completion. . . . Logically nothing more depends on them, no conclusions are drawn from them. They constitute an absolute end" (p. 222). In marking the fulfillment of scientific predictions, confirmation statements are, however, said not only to yield a characteristic satisfaction, but to influence the course of subsequent inquiry: "the hypotheses whose verification ends in them are considered to be upheld, and the formulation of more general hypotheses is sought, the guessing and search for universal laws goes on" (p. 222). The problem is whether these various features ascribed to confirmation statements can be reconciled with one another.

For, on the one hand, these statements constitute an absolute end, having no logical function when standing at the beginning of further cognitive processes, since "the moment they are gone one has at one's disposal in their place inscriptions, or memory traces, that can play only the role of hypotheses and thereby lack ultimate certainty" (p. 222). On the other hand, they enable us to uphold the hypotheses they serve to verify and to reject those they falsify, in either case leading us to conduct subsequent inquiry in a significantly different manner. If, however, a confirmation statement truly constitutes an absolute end, how can it serve thus to qualify our further treatment of relevant hypotheses? Why, indeed, should a hypothesis, supposedly verified by a confirmation statement a moment before the last, be considered *now* to have been clearly upheld, leading us to search for broader hypotheses rather than simply to continue testing the original one? We now have, after all, only the fallible record of an alleged earlier verification and, as Schlick remarks, "it is a mere assumption that it is true" (p. 220), that it agrees with its parent confirmation statement. Similarly, why should the present protocol statement recording an alleged past falsification be taken as the trace of an absolute falsification by its parent confirmation statement, since it is itself no more than a hypothesis, subject only to the weak demands of coherence? In short, if the door closed by a given confirmation statement is indeed immediately reopened, this statement can constitute no absolute end; if, on the other hand, the door remains shut, the statement clearly has a logical bearing, in fact, an unwarranted logical bearing, upon subsequent investigation. Confirmation statements, it seems, cannot bring testing processes to absolute completion without qualifying further inquiry in a manner precluded by their momentary duration. However, unless they do bring such processes to absolute completion, they have, on Schlick's account, no function at all in the economy of science. The conclusion that Schlick's account of these statements is self-contradictory seems inescapable.

The notion that confirmation statements can have no logical function for subsequent cognitive processes rests on their radical immediacy, that is, on the idea that their function "lies in the immediate present" (p. 222). Schlick thus emphasizes their differentiation from the protocol statements to which they may give rise, statements which are "always characterized by uncertainty" (p. 226). To write down a confirmation statement or even to preserve it in memory is, strictly speaking, impossible, for the meaning of critical demonstratives is altered by preservation; replacement of these demonstratives "by an indication of time and place" moreover inevitably results in the creation of "a protocol statement which as such has a wholly different nature" (p. 226). But immediacy, one may feel, should cut both ways: if it eliminates logical bearing on subsequent processes, it must equally eliminate such bearing on earlier ones. Yet Schlick holds, as we have seen, that confirmation statements bring testing processes to an absolute completion.

> Have our predictions actually come true? In every single case of verification or falsification a "confirmation" [confirmation statement] answers unambiguously with a yes or a no, with joy of fulfilment or disappointment. The confirmations are final [p. 223].

How can this be? The prediction is, after all, a scientific hypothesis with "a wholly different nature" from that of the confirmation statement in question. How can it derive any benefit from the latter's certainty any more than a later protocol statement can?

Schlick gives, as an example of a prediction: "If at such and such a time you look through a telescope adjusted in such and such a manner you will see a point of light (a star) in coincidence with a black mark (cross wires)" (p. 221). Suppose we now have the confirmation statement, "Here now a point of light in coincidence with a black mark." For the sake of argument, let us grant that the latter statement is, at the critical moment, certain. Does it follow that it constitutes an unambiguous and final answer to the question of whether the prediction has in fact come true? Not at all. For the prediction stipulates, in its antecedent clause, certain conditions relating to physical apparatus, time, and the activity of an observer. Unless the experience reported by the confirmation statement is assumed to have occurred in accordance with the conditions thus stipulated, it cannot even be judged relevant to the prediction, much less to fulfill it with finality. On the other hand, if the assumption is made that these conditions have been satisfied in fact, this critical assumption itself shares in the uncertainty of the prediction, being itself clearly no more than a physical hypothesis. The question of whether a prediction has in fact come true is, then, just the question of whether a corrigible scientific statement, rather than a confirmation statement, is true. Such a question can always be reopened. The conclusion must be that the alleged certainty of confirmation statements no more enables them to provide absolutely certain completions for earlier scientific processes than it equips them to constitute absolute origins.

The alleged certainty of these statements must, finally, be called into question: What does such certainty amount to? According to Schlick, I cannot be deceived regarding the truth of my own confirmation statements, even though, as he writes, "the possibilities of error are innumerable" (p. 212). He admits, of course, that protocol statements are subject to error. Indeed, in one passage he concedes that a protocol statement saying "that N.N. used such and such an instrument to make such and such an observation" may be mistaken, because N.N. may "inadvertently . . . have described something that does not accurately represent the observed fact" (p. 212). Does this not imply that, in such a case, N.N. was himself deceived regarding the truth of his own confirmation statement? Such an implication is certainly unwelcome to Schlick, and contrary to his fundamental view. The question remains as to how it might reasonably be avoided. What is it, indeed, that precludes errors due to inadvertence and poor judgment in the case of confirmation statements alone, allowing them full scope in all other cases?

Schlick certainly does not suppose that one's own statements are generally immune to error. "Even in the case of statements which we ourselves have put forward," he writes, "we do not in principle exclude the possibility of error" (p. 213). However, he continues immediately to illustrate his point with specific and exclusive reference to protocol statements: "We grant that our mind at the moment the judgment was made may have been wholly confused, and that an experience which we now say we had two minutes ago may upon later examination be found to have been a hallucination, or even one that never took place at all" (p. 213). A statement by which we affirm the occurrence of our experience of two minutes earlier can, of course, in his scheme, not be a confirmation statement at all but only a protocol statement. The question persists, however: Is mental confusion possible only in the making of protocol statements? What protects us from confusion in the judgment of simultaneous experience? What necessity ensures us against error?

Concerning protocol statements occasioned by confirmation statements, Schlick writes that they are to be classed as hypotheses. "For, when we have such a statement before us, it is a mere assumption that it is true, that it agrees with the observation statements [confirmation statements] that give rise to it" (p. 220). The supposition is evident that if a protocol statement disagrees with the relevant confirmation statement, it must be the former and not the latter which is false. But why this asymmetry? What prevents the protocol statement from giving a truer account of a man's experience at a given moment than his own description at that moment? In one passage, Schlick suggests the curious doctrine that the protocol statement reports not a perception or an experience but rather the occurrence of a confirmation statement. The protocol statement, "M.S. perceived blue on the nth of April 1934 at such and such a time and such and such a place," he declares equivalent to "M.S. made . . . (here time and place are to be given) the confirmation 'here now blue' " (p. 226). Under such a doctrine, to be sure, if M.S.'s actual confirmation statement at the relevant time and place was "Here now yellow," the protocol statement must be false. It does not follow, however, that the confirmation statement is true. And the proposed equivalence must itself be rejected, for to perceive blue and to say "Here now blue" are surely different. If it is false that M.S. *said* "Here now blue," it is not therefore false that he *perceived* blue. Despite the fact that he actually said "Here now yellow," he may have perceived blue: his confirmation statement itself may have been false. What eliminates such a possibility?

The fundamental answer, in Schlick's theory, is given by his reference to the demonstrative terms included in confirmation statements. As he puts it, " 'This here' has meaning only in connection with a gesture" (p. 225). To comprehend the meaning of a confirmation statement, "one must somehow point to reality" (p. 225). It follows, in his view, that I cannot understand a confirmation statement without thereby determining it to be true. Here is the fundamental source of the certainty of confirmation statements: their understanding presupposes their verification. Now it may perhaps be argued that the certainty thus yielded is weak because it is dependent upon the notion of "understanding." (I may, indeed, not be at all sure I understand a given confirmation statement when I make it, and may later decide that in fact I did not.) But the trouble, in any case, runs much deeper than this. Schlick's basic conception of the matter rests, I believe, upon a confusion.

For suppose it be granted that the meaning of demonstrative terms derives from their function as gestures, by which, as Schlick remarks, "the attention is directed upon something observed" (p. 225). Suppose it be admitted that "in order therefore to understand the meaning" of a confirmation statement, "one must simultaneously execute the gesture, one must somehow point to reality" (p. 225). What can be inferred from such admissions? They imply only that the comprehension of a confirmation statement requires attention to those observed elements indicated by its constituent demonstrative terms. In this and this sense only can comprehension of the statement be said to involve a "pointing to reality." By no means is it implied that we must point to reality in the wholly different sense of verifying the attribution represented by the statement as a whole. To attend to an indicated thing cannot be equated with determining the truth of any affirmation concerning it. It is therefore simply fallacious to infer, from the presumed necessity of attending to the things indicated by embedded demonstratives, the necessity of verifying the assertion made by a statement with the help of such demonstratives, in order to grasp its meaning.

The fallacy is concealed by the equivocal phrase "pointing to reality." For, in explaining the latter phrase, initially introduced to denote attention to the reference of a demonstrative term functioning as a gesture, Schlick writes: "In other words: I can understand the meaning of a 'confirmation' [confirmation statement] only by, and when, comparing it with the facts, thus carrying out that process which is necessary for the verification of all synthetic statements" (p. 225). Here is the easy, but illegitimate shift from the reference of a term to the reference of a statement, from things indicated to facts expressed, from attending to the object of a demonstrative to verifying the truth of a statement. Once this shift is exposed, Schlick's fundamental argument for the certainty of confirmation statements falls to the ground: the understanding of such statements does not, after all, presuppose their verification. I may understand a confirmation statement and be undecided as to its truth; what is more, I can understand, and even affirm, a confirmation statement that is false. Schlick's positive theory, no less than Neurath's, thus proves untenable.

The failure of both these theories may engender despair. For they seem, between them, to exhaust the possibilities for dealing with our basic dilemma between coherence and certainty. Either some of our beliefs must be transparently true of reality and beyond the scope of error and revision, or else we are free to choose any consistent set of beliefs whatever as our own, and to define "correctness" or "truth" accordingly. Either we suppose our beliefs to reflect the facts, in which case we beg the very question of truth and project our language gratuitously upon the world, or else we abandon altogether the intent to describe reality, in which case our scientific efforts reduce to nothing more than a word game. We can, in sum, neither relate our beliefs to a reality beyond them, nor fail so to relate them.

Despite this grim appraisal, I believe that despair is avoidable, and that the general approach of the previous lectures can be sustained. My view is that while rejecting certainty, it is yet possible to uphold the referential import of science; that to impose effective constraints upon coherence need beg no relevant questions nor people the world with ghostly duplicates of our language. In the remainder of the present lecture, I shall attempt to spell out the reasons for these views.

Let me turn first to the fundamental opposition between coherence and certainty. We have seen how central this opposition is in the thought of both Neurath and Schlick, who take contrary positions. Neurath insists that every scientific statement is subject to change, urging that "There is no way of taking conclusively established pure protocol sentences as the starting point of the sciences" (PS, 201). Rejecting certainty, he rejects, however, also the very idea of checking beliefs against experience or reality, taking coherence to be the only defensible alternative. Schlick, on the other hand, reacting against the absurdities of coherence, is driven to seek a class of "absolutely certain" (p. 223) statements capable of providing an "unshakeable point of contact between knowledge and reality" (p. 226). Affirming the need to relate science to fact, he assumes that only a doctrine of certainty can make such a relationship intelligible. Schlick and Neurath are thus agreed in binding extra-linguistic reference firmly to certainty and they join, therefore, in reducing the effective alternatives to two: (1) a rejection of certainty, as well as of appeals to extra-linguistic reference, yielding a coherence view, and (2) a rejection of the coherence view in favor of an appeal to extra-linguistic reference, yielding a commitment to certainty. Given such a reduction, we are driven to defend coherence if we find certainty repugnant, and impelled to defend certainty if appalled by the doctrine of coherence. The reduction makes it possible for these equally unpleasant alternatives thus to feed upon each other. But this reduction itself must be rejected. There is, in fact, no need to assume that the alternatives are exhausted by coherence and certainty; a third way lies open.

For what is required is simply a steady "referential" limitation upon unbridled coherence; certainty supplies much more than is required. In particular, it imports the notion of a fixity, a freedom from error and consequent revision, which cannot be defended for it is nowhere to be found. The supposition that certainty is required to restrain coherence is perhaps in part due to an ambiguity in the notion of "fixity": in one sense, a fixed point is simply one that is selectively designated, providing us with a frame of reference or a standard, for purposes of a given sort; in another sense, a fixed point is one that, in particular, does not undergo change over time. A point may be fixed in the first sense without being fixed in the second. There may be no temporally constant or *permanently* fixed points in a given context, and yet there may be fixed points that are relevant and effective in that context, at each moment. The scope of unrestrained coherence needs, indeed, to be supplemented by the introduction of relevantly "fixed" points at all times. However, the further supposition that there must be *some* points that are permanently fixed, that is, held forever immune to subsequent revision, is simply gratuitous.

We have earlier cited Schlick's remark that the coherence theory can be avoided only if we specify statements that "are to be maintained, to which the remainder have to be accommodated" (p. 216). That certain statements need to be *maintained* over time is an unwarrantedly strong demand. We need only recognize that statements have referential values for us, independent of their consistency relationships to other statements, and that these values, though subject to variation over time, provide us, at each moment, with sufficient "fixity" to constitute a frame of reference for choice of hypotheses.

How are these values of statements, compatible with lack of certainty, to be conceived? They may be thought of as representing our varied inclinations to affirm given statements as true or assert them as scientifically acceptable; equivalently, they may be construed as indicating the initial claims we recognize statements to make upon us, at any given time, for inclusion within our cognitive systems. A notion of this general sort has been put forward by Bertrand Russell in *Human Knowledge,* under the label "intrinsic credibility,"[8] and

---

[8] Bertrand Russell, *Human Knowledge* (New York: Simon and Schuster, 1948), Part II, chap. 11, and Part V, chaps. 6 and 7.

Nelson Goodman has spoken, analogously, of "initial credibility,"[9] the adjective serving in each case to differentiate the idea in question from the purely relative concept of "probability with respect to certain other statements." Goodman explains his conception as follows:

> Internal coherence is obviously a necessary but not a sufficient condition for the truth of a system; for we need also some means of choosing between equally tight systems that are incompatible with each other. There must be a tie to fact through, it is contended, some immediately certain statements. Otherwise compatibility with a system is not even a probable indication of the truth of any statement.
>
> Now clearly we cannot suppose that statements derive their credibility from other statements without ever bringing this string of statements to earth. Credibility may be transmitted from one statement to another through deductive or probability connections; but credibility does not spring from these connections by spontaneous generation. Somewhere along the line some statements, whether atomic sense reports or the entire system or something in between, must have initial credibility. So far the argument is sound. . . . Yet all that is indicated is credibility to some degree, not certainty. To say that some statements must be initially credible if any statement is ever to be credible at all is not to say that any statement is immune to withdrawal. For indeed, . . . no matter how strong its initial claim to preservation may be, a statement will be dropped if its retention—along with consequent adjustments in the interest of coherence—results in a system that does not satisfy as well as possible the totality of claims presented by all relevant statements. In the "search for truth" we deal with the clamoring demands of conflicting statements by trying, so to speak, to realize the greatest happiness of the greatest number of them. These demands constitute a different factor from coherence, the wanted means of choosing between different systems, the missing link with fact; yet none is so strong that it may not be denied. That we have probable knowledge, then, implies no certainty but only initial credibility.[10]

There is much in the above account that is metaphorical, but the fundamental point for present purposes seems to me quite simple

[9] Nelson Goodman, "Sense and Certainty," *Philosophical Review,* LXI (1952), 160–167.
[10] *Ibid.,* pp. 162–163.

and persuasive: while certainty is untenable, it is also excessive as a restraint upon coherence. Such restraint does not require that any of the sentences we affirm be guaranteed to be forever immune to revision; it is enough that we find ourselves now impelled, in varying degrees, to affirm and retain them, seeking to satisfy as best we can the current demands of all. That these current demands vary for different, though equally consistent statements, and that we can distinguish, even roughly, the credibility-preserving properties of alternative coherent systems, suffices to introduce a significant limitation upon coherence. For it means that not all coherent systems are equally acceptable; we are not free to make an arbitrary choice among them. Nor, where what has hitherto been a satisfying system conflicts with a strong new candidate for inclusion, are we free to decide the matter arbitrarily, discarding one or the other at will. There is a price to be paid in either case in terms of overall credibility-preservation and, though it may be difficult to determine in given circumstances, it is a clearly relevant consideration.

In any event, it is the claims of sentences at a given time which set the problem of systematic adjudication at that time, and so restrain the arbitrariness of coherence. That these claims may vary in the future does not alter the present task. That a sentence may be given up at a later time does not mean that its present claim upon us may be blithely disregarded. The idea that once a statement is acknowledged as theoretically revisable, it can carry no cognitive weight at all, is no more plausible than the suggestion that a man loses his vote as soon as it is seen that the rules make it possible for him to be outvoted.

There is, to be sure, a certain awkwardness in expressing a statement's theoretical revisability in the same breath with asserting the statement itself, and the difficulty this presents may be one source of the deep-rooted feeling that certainty, rather than credibility, is required. If I say "There's a horse but it may not be", I may appear to be at odds with myself in a peculiar way; to remove the conflict I must presumably assert "There's a horse" unqualifiedly and defend the immunity of this statement to possible withdrawal at any future time. Such a cure would, however, be worse than the disease, for by this argument all statements would be certain. The problem is more easily handled by finding a suitably circumspect form of interpretation, e.g., "There's a horse; and I recognize the possibility that the statement just made might be withdrawn under other circumstances than those now prevailing". The recognition of the latter possibility clearly does not conflict with the initial assertion; such possibility surely does not need to be eliminated in order to make the way clear for assertion generally.

It follows that the basic dilemma with which we started, between coherence and certainty, collapses. That none of the statements we assert can be freed of the possibility of withdrawal does not imply that no statement exercises any referential constraint at any time. That none can be *guaranteed* to be an absolutely reliable link to reality does not mean that we are free to assert any statements at will, provided only that they cohere. That the statement "There's a horse" cannot be rendered theoretically certain does not permit me to call anything a horse if only I do not thereby contradict any other statement of mine. On the contrary, if I have learned the term "horse", I have acquired distinctive habits of individuation and classification associated with it; I have learned what Quine refers to as its "built-in mode . . . of dividing [its] reference."[11] These habits do not guarantee that I will never be mistaken in applying the term, but it by no means therefore follows that they do not represent selective constraints upon my mode of employing the term. On the contrary, such constraints generate credibility claims which enter my reckoning critically as I survey my system of beliefs. I seek not consistency alone, but am bound to consider also the relative inclusiveness with which a system honors initial credibilities.

It follows, therefore, that the emphasis of earlier lectures on the *independence* of observation statements as the primary locus of their control does not, after all, have the consequence of committing us to the coherence theory. Such statements, as we urged, are not isolated certainties, but must be accommodated with other beliefs in a process during which they may themselves be overridden. It

[11]Willard Van Orman Quine, *Word and Object* (New York and London: The Technology Press of the Massachusetts Institute of Technology, and John Wiley & Sons, Inc., 1960), p. 91.

is, however, enough for the purposes of control that they may clash with these other beliefs in such a way as to force an unsettling systematic review of the situation. Such a review is not (contrary to the hypothetical fears earlier expressed) motivated simply by the wish to restore consistency. The need is, of course, to maintain consistency, but also to sacrifice as little as possible of overall credibility.

An observational expectation induced in us by our heretofore satisfying system may, for example, be challenged by an experimental observation which drastically increases the credibility of a statement incompatible with this expectation while radically reducing the credibility of the expectation itself. The problem in such a situation is to determine which consistent alternative strikes a more inclusive balance of relevant credibility claims. To drop the initial expectation in favor of the more credible incompatible statement demands internal systematic revision in the interests of consistency. To exclude the incompatible statement and maintain the system intact lowers the overall credibility value of the latter, for the credibility loss of its constituent expectation reverberates inward. Every clash-resolution, in short, has its price. In some such situations, the choice may be relatively easy; in others it may be exceedingly delicate; and it may even, in some circumstances, defy resolution.

Nevertheless, it is clear that we are in no case free simply to choose at will among all coherent systems whatever. And, further, it is clear that the control exercised by observation statements does not hinge on certainty. It requires only that the credibility they acquire at particular times be capable of challenging, in the manner above described, the expectations flowing from other sources. Such control is, surely, not absolute, since any observation statement may itself be outvoted in the end; in pressing their independent claims, however, such statements nevertheless contribute, along with other statements, to the restraint of arbitrariness in the choice of beliefs. Control is, moreover, released from distinctive ties to any special sort of statement, and diffused throughout the realm of statements as a whole.

What shall now be said concerning the difficult notions of truth and reality? Eschewing certainty, some philosophers, Neurath included, have rejected all talk of reality and truth. Having pointed out that appeal to an immediate comparison with the facts as a method of ascertaining truth is question-begging, and that facts, construed literally as entities, are mere ghostly doubles of true sentences, they have proceeded to cast doubt upon all thought of external reference, as embodied in philosophically innocent talk of reality and fact, and in innocent as well as serious talk of truth. Such skepticism leads, however, to insuperable difficulties, for without external reference, science has no point. If we stay within the circle of statements altogether, we are trapped in a game of words, with which even Neurath (as indicated by his reference to science as an instrument for life) cannot be wholly satisfied. Taken in its extreme form, Neurath's doctrine understandably evokes the sort of criticism which Russell offers:

> Neurath's doctrine, if taken seriously, deprives empirical propositions of all meaning. When I say 'the sun is shining', I do not mean that this is one of a number of sentences among which there is no contradiction; I mean something which is not verbal, and for the sake of which such words as 'sun' and 'shining' were invented. The purpose of words, though philosophers seem to forget this simple fact, is to deal with matters other than words. . . . The verbalist theories of some modern philosophers forget the homely practical purposes of every-day words, and lose themselves in a neo-neo-Platonic mysticism. I seem to hear them saying 'in the beginning was the Word,' not 'in the beginning was what the word means.' It is remarkable that this reversion to ancient metaphysics should have occurred in the attempt to be ultra-empirical.[12]

One source of the trouble is a persistent confusion between truth and estimation of the truth, between the import of our statements and the processes by which we choose among them. If, for example, appeal to reality or direct comparison with the facts is defective as a method of *ascertaining* truth, this does not show that the *purport* of a true statement cannot properly be described, in ordinary language, as "to describe reality" or "to state the facts." We may have

[12]Bertrand Russell, *An Inquiry into Meaning and Truth* (London: Allen and Unwin, 1940), pp. 148–149; Penguin edn. (Harmondsworth: Penguin Books, Ltd., 1962), pp. 140–141.

no certain intuition of the truth, but this does not mean that our statements do not purport to be true. Now, if the sentence "Snow is white" is true, then snow is (really, or in fact) white, and vice versa, as Tarski insists.[13] As Quine has remarked, "Attribution of truth in particular to 'Snow is white', for example, is every bit as clear to us as attribution of whiteness to snow."[14] We may be unclear as to how to decide whether the sentence "Snow is white" is true, but the sentence in any case *refers* to snow and claims it to be white, and if we decide to hold the sentence to be true, we must be ready to hold snow to be (really, or in fact) white. The question of how we go about deciding what system of descriptive statements to accept in science is the question of how to estimate what statements are true. Whatever method we employ, our statements will refer to things quite generally and will purport to attribute to them what, *in reality, or in fact,* is attributable to them.

There is thus no way of staying wholly within the circle of statements, for in the very process of deciding which of these to affirm as true, we are deciding how to refer to, and describe things, quite generally. The *import* of our statements is inexorably referential. It is, however, quite another matter to suppose that, because this is so, the *methodology* by which we accept statements may be philosophically described in terms of an appeal to such supposed entities as facts, with which candidate statements are to be directly compared. For facts, postulated as special entities corresponding to truths, are generally suspect, and the determination of their existence is question-begging if proposed literally as a method of ascertaining the truth.

We thus separate the question of the *import* of scientific systems from the question of the *methods* by which we choose such systems. Can such methods be described without dependence upon the no-

[13] Alfred Tarski, "The Semantic Conception of Truth," *Philosophy and Phenomenological Research* (1944); reprinted in Herbert Feigl and Wilfrid Sellars, eds., *Readings in Philosophical Analysis* (New York: Appleton-Century-Crofts, Inc., 1949), pp. 52–84.

[14] Willard Van Orman Quine, *From A Logical Point of View* (Cambridge, Mass.: Harvard University Press, 1953), p. 138.

tion of direct comparison with the facts? Both Neurath and Schlick assume that any conception of science as referential must display such dependence. This supposition, however, is simply false. The conception of credibility above sketched represents a notion of choice among systems of statements, yet makes no use of any idea such as that of comparison with the facts. It is, nevertheless, perfectly compatible with the recognition that any system selected is referential in its *import*. Moreover, credibility considerations rest on the referential values which statements have for us at a given time, that is, on the inclinations we have, at that time, to affirm these statements as true.

Such inclinations as to statements are, surely, tempered by habits of individuation and classification acquired through the social process of learning our particular vocabulary of *terms*. In learning the term "horse", for example, I have incorporated selective habits of applying and withholding the term; these habits, operating upon what is before me, incline me to a greater or lesser degree to affirm the statement "There's a horse". If I have learned the term "white" as well as "horse," I may, further, be strongly inclined, on a given occasion, to affirm "That horse is white", and my inclination, on such an occasion, will be understandable in part, as a product of my applying both "horse" and "white" to what I see. I need, however, surely not recognize any such additional entity as *the fact that that horse is white,* nor need I have an applicable term for such a supposed entity in my vocabulary. Though it hinges in various ways on referential habits associated with a given vocabulary, the notion of initial credibility thus requires no reference to *facts,* in particular.

To be sure, whether I am to accept a statement depends not only on my initial inclination to accept it, but rather on its fitting coherently within a system of beliefs that is sufficiently preserving of relevant credibilities. Here again, however, there is no reference to any such entities as facts, with which statements are to be compared. In accepting a system I nevertheless take its *import* to be referential: I hold its statements to be true, and genuinely accept whatever attributions it makes to the entities mentioned in these statements. In a philosophically harmless sense, I may then say that

I take the system as expressive of the facts. I have, at no time, any guarantees that my system will stand the test of the future, but the continual task of present evaluation is the only task it is possible for me to undertake. Science, generally, prospers not through seeking impossible guarantees, but through striving to systematize credibly a continuously expanding experience.

Appendix A

Vision and Revolution: A Postscript on Kuhn[*]

Introduction. In Chapter 4 of *Science and Subjectivity,* I offered several arguments critical of Professor Thomas Kuhn's views as expressed in his influential book *The Structure of Scientific Revolutions.*[1] His recent replies to these criticisms[2] seem to me so inadequate as to suggest that he, and therefore others as well, may have failed to grasp their full import. Accordingly, I shall, in the first part of this paper, briefly recapitulate my earlier arguments and offer a short rejoinder to Professor Kuhn's replies.[3] The second part of the paper will expand upon my earlier discussion to consider the basic metaphors of *vision* and *revolution,* offered by Kuhn to replace the traditional notion of deliberation. My argument here will be that these new metaphors are incongruous in critical respects, and my discussion will conclude by considering their relations to the contrast between understanding and accepting a theory.

* Reprinted from *Philosophy of Science,* vol. 39 (1972), pp. 366–374, by permission of the publisher.

[1] Thomas S. Kuhn, *The Structure of Scientific Revolutions* (Chicago and London: The University of Chicago Press, 1962). Because of the numerous citations in the present paper, page references to passages in *The Structure of Scientific Revolutions* will hereafter be given directly following quoted portions in the text, enclosed within parentheses. Note that all such references are to the first edition of 1962, rather than to the second edition of 1970.

[2] In the second edition of *The Structure of Scientific Revolutions,* 1970, "Postscript-1969," pp. 174–210; and in "Reflections on my Critics" in I. Lakatos and A. Musgrave, eds., *Criticism and the Growth of Knowledge* (Cambridge: Cambridge University Press, 1970), pp. 231–278).

[3] I do not offer a comprehensive examination of Kuhn's replies to his critics, however. Dudley Shapere, in "The Paradigm Concept," *Science* 172 (1971): 706 ff., reviews the general effect of Kuhn's recent replies in a discussion I find persuasive.

Part I: Recapitulation and Rejoinder

In criticizing Professor Kuhn's view, my discussion in *Science and Subjectivity*[4] offered the following arguments:

(1) *Self-refutation:* If paradigm debates are characterized by an "incompleteness of logical contact" (p. 109) between proponents of rival paradigms, and the transition to a new paradigm does not occur "by deliberation and interpretation" (p. 121), it is self-defeating to justify this view itself by deliberation appealing to factual evidence from the history of science. If it is true that "paradigm changes do cause scientists to see the world of their research-engagement differently . . . that after a revolution scientists are responding to a different world" (p. 110), there can be no appeal to ostensibly paradigm-neutral factual evidence from history in support of Kuhn's own new paradigm. Conversely, if *historians* can transcend particular paradigms and evaluate them by appeal to neutral evidence, so can *scientists,* i.e., they can engage in rational paradigm debates which are perfectly intelligible. [See *SS*, 21–22, 53, 74]

(2) *Observational difference:* It does not follow, from the fact that different paradigms organize their observations differently, that they are directed to different objects—that "after a revolution scientists are responding to a different world" (p. 110). From the fact that certain items are seen in varying ways under diverse categorizations, it cannot be inferred that they are not identical. There is a contrast between seeing x and seeing x *as* something or other. [See *SS*, 41]

(3) *Meaning variation:* Paradigm change does not inevitably alter constituent meanings through altering the language or definitions of basic terms, with the result that "communication across the revolutionary divide is inevitably partial" (p. 148). The supposition of *inevitable* meaning change confuses the notion of a language as consisting simply of a vocabulary and grammar, with that of a language construed as a system of assertions. Or else it overlooks the fact that variation of definition or sense is consistent with referential stability and

---

[4] Hereafter, page references to this book will be given directly in the text, prefixed by the letters *"SS,"* and enclosed in parentheses. (Note that *parenthesized* page references *without* the prefix *"SS" always* refer to Kuhn's first edition, as explained above in footnote 1.)

that only the latter is required for the stability of deduction. [See *SS*, 58–62, 64]

(4) *Gestalt switch:* Assuming that adoption of a paradigm is accompanied by a "relatively sudden" and "intuitive" alteration of perception such as the "gestalt switch," it does not follow that there are no public procedures of assessment by which such a paradigm is evaluated after it is originated. If, as Kuhn says, "no ordinary sense of the term 'interpretation' fits these flashes of intuition through which a new paradigm is born" (p. 122), it may not be inferred that the term 'interpretation' does not apply to the processes by which the scientific community debates the merits of the new paradigm. [See *SS*, 78–79]

(5) *Perception vs. debate:* If intuitive processes of perception characterize the psychology of the paradigm originator, he himself does not defend his paradigm by simple appeal to such processes. He engages in debate and proposes arguments. If he is deluded about the force of these arguments, the delusion cannot be demonstrated simply by arguing from the psychology of perception, but requires appeal to the content of the relevant debates. Nor are *particular examples* of debates at cross-purposes sufficient. For common criteria may have borderline regions of indeterminacy, and even where determinate, there may well be differences of judgment in application, as well as misunderstanding. [See *SS*, 79–80]

(6) *Inadequacy of normal science:* Paradigm choice cannot be resolved by normal science, according to Kuhn, for if differing scientific schools disagree about the character of problems and solutions, "they will inevitably talk through each other when debating the relative merits of their respective paradigms" (p. 108). Thus, deliberation gives way to persuasion and conversion. But to assume that deliberation and interpretation are *restricted* to normal science begs the very question at issue. "If paradigms are not corrigible by normal science, it does not follow that they are not corrigible at all. If scientific schools inevitably talk through each other when arguing from within their respective paradigms, it is not further inevitable that they do always argue from within their respective paradigms." [*SS*, 80–81]

(7) *Incommensurability of paradigms:* Competing paradigms, for Kuhn, are addressed to different problems, embody different standards and different definitions of science (p. 147); they are

based on different meanings and operate in different worlds (pp. 148–149) [*SS*, 81]. Therefore, he argues, "The proponents of competing paradigms are always at least slightly at cross-purposes" (p. 147). Before they can communicate, "one group or the other must experience the conversion that we have been calling a paradigm shift. Just because it is a transition between incommensurables, the transition . . . cannot be made a step at a time, forced by logic and neutral experience" (p. 149). But if the paradigms are indeed so different how can they be in competition? If they are indeed rivals, they must be accessible to some shared perspective within which they can be compared. Incommensurability does not imply incomparability. [See *SS*, 82–83]

(8) *Second-order incommensurability:* Proponents of different paradigms, says Kuhn, acquire different criteria for determining relevant problems and solutions. The debate between them is circular or inconclusive because "each paradigm will be shown to satisfy more or less the criteria that it dictates for itself and to fall short of a few of those dictated by its opponent" (pp. 108–109). "Paradigm differences are thus inevitably reflected upward, in criterial differences at the second level" [*SS*, 84]. This argument, however, confuses *internal* criteria, by which paradigms determine problems and solutions, with *external* criteria by which they are themselves judged. The latter are independent of the former, and, hence, the argument that paradigms must inevitably be self-justifying collapses. [See *SS*, 84–86]

(9) *Resistance to falsification:* Paradigms are not rejected when counterinstances occur, until an alternative is available, argues Kuhn, while acceptance of a new and untried paradigm depends on faith (pp. 77, 157). Thus, he concludes "the competition between paradigms is not the sort of battle that can be resolved by proofs" (p. 147). But proof is in any case irrelevant, and faith in a new hypothesis is compatible with acknowledgement of shared procedures of evaluation by which the hypothesis is to be assessed. Nor is loss of faith in the truth of a hypothesis inconsistent with its continued use as a practical tool of research or application, or with belief that some modification of the hypothesis is true. Kuhn himself seems to admit the critical point by allowing that, although scientists do not renounce the paradigm provoking crisis, they "may begin

to lose faith and then to consider alternatives" (p. 77). The global use of terms such as 'acceptance', 'rejection' and 'paradigm' itself, prevents his full appreciation of this point. [See *SS*, 86–88]

(10) *Cumulativeness:* In denying the cumulative character of scientific change, Kuhn says "the successful new theory must somewhere permit predictions that are different from those derived from its predecessor. That difference could not occur if the two were logically compatible" (p. 96). But if the two are incompatible logically, they must be, at least in part, commensurable, and *a fortiori*, comparable. It follows also that accumulation, while not necessary, is at least logically possible. Further, despite his criticism of the notion of cumulative science, he says that "new paradigms . . . usually preserve a great deal of the most concrete parts of past achievement and they always permit additional concrete problem-solutions besides" (p. 168). [See *SS*, 89, 65]

(11) *Reintroduction of criticized notions:* Notions criticized by Kuhn reemerge often under new labels, in his theory. Thus, as just noted, commensurability is implied by his emphasis on logical incompatibility. Falsification returns under the guise of anomaly, crisis, and loss of faith. Cumulativeness is acknowledged in the preservation and extension of past achievement, as noted. Interpretation and deliberation are acknowledged in all the above ways as well as his emphasis on the promise of a new paradigm "to resolve some outstanding and generally recognized problem that can be met in no other way" (p. 168). The critical distinction between theory-genesis and theory-justification is thus, in effect, reinstated. (If we take Professor Kuhn's denials alone, they form a radical and interesting departure from older views, but they seem clearly untenable. If we take just his affirmations of older concepts under new labels, we have a plausible but no longer novel view. If we take both the denials and the affirmations, we find an inconsistent account. The strong and exciting part of this account can be defended by judicious retreat to the weaker and unexciting part, but in no case do we have both a radical and a tenable view of science.) [See *SS*, 88–89]

In his recent replies, Professor Kuhn responds to my criticisms and those of others by insisting that his book does contain "a pre-

liminary codification of good reasons for theory choice. These are
. . . reasons of exactly the kind standard in philosophy of science:
accuracy, scope, simplicity, fruitfulness, and the like."[5] They are,
however, he insists, "values" rather than "rules of choice," and
may be applied differently by different scientists.[6] Nowhere, however, have I argued that there must be rules *applied uniformly* by
scientists. If these reasons are not, in being conceived as *values,*
to be construed as utterly free of all constraints or themselves
paradigm-dependent, then they allow the comparison of rival paradigms and make paradigm debates intelligible. Such intelligibility
is, however, at odds with Kuhn's statement that the proponents of
competing paradigms "will inevitably talk through each other"
(p. 108).

In his *Postscript,* he refers to my discussion along with others'
as follows: "Because I insist that what scientists share is not sufficient to command uniform assent about such matters as the choice
between competing theories or the distinction between an ordinary anomaly and a crisis-provoking one, I am occasionally ac-
cused of glorifying subjectivity and even irrationality."[7] Now I
cannot be certain what others have said but it is clear that my
arguments above enumerated do not include a demand for uniform
assent. Indeed, I explicitly stated [*SS,* 80], that "the existence of
common evaluative criteria is compatible with borderline regions
in which these criteria can yield no clear decisions. And . . . even
the objective availability of clear decisions is consistent with honest differences of judgment, not to mention plain misunderstandings." The issue is not *uniformity* but *objectivity,* and objectivity
requires simply the possibility of intelligible debate over the comparative merits of rival paradigms.[8]

---

[5] I. Lakatos and A. Musgrave, eds., *Criticism and the Growth of Knowledge,*
op. cit., p. 261.

[6] *Ibid.,* p. 262.

[7] *The Structure of Scientific Revolutions,* Second edition, p. 186.

[8] For a discussion of Kuhn's replies relating also to other issues, see Shapere's
review "The Paradigm Concept", *op. cit.,* (footnote 3 above).

Part II: Vision and Revolution

I turn now to a consideration of certain broader aspects of Professor Kuhn's treatment, which I have not previously discussed in any detail. My earlier critique does note the "striking way in which Kuhn's account applies psychological, political, and religious categories to the description of scientific change. The older references to logical system, observational evidence, theoretical simplicity, and experimental test have given way, in his account, to mention of the gestalt switch, conversion, faith, decision, and death" [*SS*, 78]. We have a replacement of the categories of interpretation and deliberation by new categories or metaphors in treating of science. The two fundamental and controlling metaphors are those of *vision* and *revolution*. What I shall now argue is that these are incongruous with one another in a philosophically significant way. Neither, taken alone, supports the view of paradigm change offered by Kuhn to replace the traditional conception. What supports Kuhn's view is rather a hybrid that resembles neither vision nor revolution, but is violated by both. This question seems to me of general interest, independently of the particulars of arguments earlier reviewed, for what is at stake is the fundamental aspect under which we attempt to make science intelligible to ourselves, the controlling analogies by which we tend to think of it.

Take first the topic of *vision*. Kuhn assimilates theoretical change to a gestalt reorganization of vision. "What were ducks in the scientists' world before the revolution are rabbits afterwards. The man who first saw the exterior of the box from above later sees its interior from below" (p. 110). The patterned alteration of thought and experience concomitant with theoretical change is plausibly compared with the intuitive and spontaneous shift in perception of a reversible figure, and resembles it better than it does the piecemeal articulations associated with deliberation. Anomalies and crises, says Kuhn, "are terminated, not by deliberation and interpretation, but by a relatively sudden and unstructured event like the gestalt switch. Scientists then often speak of the 'scales falling from the eyes' or of the 'lightning flash' that 'inundates' a previously obscure puzzle, enabling its components to be seen in a new way that for the first time permits its solution. . . . No ordinary sense of the term 'interpretation' fits these flashes of intuition through which a new paradigm is born" (pp. 121–122).

The metaphor of *revolution* is employed in *conjunction* with that of the gestalt switch. "What were ducks . . . before the revolution are rabbits afterwards" (p. 110). But the notion of revolution is ramified in further ways. Competition is visualized as combat, with victory the prize. The conflict is a matter of "techniques of persuasion, or about argument and counterargument in a situation in which there can be no proof" (p. 151). Progress always accompanies victory because the winning camp is in a position to rewrite the textbooks and the implicit history of the subject. And Kuhn cites Planck's statement that "a new scientific truth does not triumph by convincing its opponents and making them see the light, but rather because its opponents eventually die, and a new generation grows up that is familiar with it" (p. 150).

Now a closer look at these two controlling metaphors reveals a critical incongruity between them. While there is, both in the case of the reversible figure and the case of revolutionary conflict a certain mutual exclusiveness of elements, the notion of an opposition of *claims* applies only to the latter. Consider the two views of the reversible cube: These two views are, as a matter of natural fact, exclusive at any given moment. We can flip from one to the other and may even come to shuttle back and forth between them with a certain amount of familiarity. However, they cannot both be seen simultaneously. Such exclusivity may be superficially assimilated to that of the revolutionary situation in which our allegiance must, at any time, be given to one side or the other—where we may shift loyalties but where it is impossible to be loyal to both sides at once or to join segments of each in some form of compromise. However—and this is the critical point—a revolution is a matter of opposed *loyalties and allegiances,* of conflicting *judgments and claims,* whereas there are no analogous questions of loyalty or allegiance or of conflicting claims in the case of alternative views of a reversible figure.

Accordingly, having appreciated the reversibility of the duck-rabbit, there is no question of *arguing* over the relative merits of the duck or the rabbit as the *proper and exclusive* view of the duck-rabbit figure, nor is there any clash of opposed loyalties in the interpretation of the reversible cube. No one *advocates* exclusive acceptance of the exterior-superior view of the cube as better than the interior-inferior view, nor is there any question of deliberate assent or commitment to one, rather than the other. As there are no arguments over the exclusive merits of the duck or the rabbit, there is no duck-party seeking victory over the rabbit-party or aiming to consign it to the dustbin of history. Nothing in this situation indeed corresponds to paradigm debates in the case of theoretical controversies. Nor are there anything like "good reasons" offered in such controversies.

The case of revolution is quite different. Each side seeks victory, demands exclusive allegiance, claims superiority, expresses commitment, propounds arguments, engages in interpretation and persuasion, formulates its rationales, rebuts the arguments of the opposition. Nor is each party totally enclosed in its own conceptual and rhetorical box. It expresses its own view, to be sure, but it attacks the views of its opponents, claiming to understand them well enough to refute them. It addresses its arguments not only to its convinced adherents and to susceptible opponents but to as-yet neutral bystanders whose allegiances are sought, and to whose thought-world the revolutionary line needs to be made to relate. Revolutionary argument is, to be sure, not necessarily scientific in spirit: it may be dogmatic, metaphysical, deceptive or otherwise defective from a scientific point of view. However, it claims commitment, it expresses advocacy, it offers arguments, it gives reasons, it propounds interpretations, it demands acceptance. To reduce the combat of revolutionary parties to a gestalt switch is to *leave out* the critical aspect of *advocacy and opposed loyalties;* it is to omit the notion of a *claim* and that of a *rationale.* Whereas, conversely, to offer the gestalt switch as a case of revolutionary transition is to *import* the inapplicable concepts of advocacy, commitment, and party combat to merely phenomenally alternative perceptual configurations.

The latter point is particularly important. For if we imagine that the relatively spontaneous intuitive process by which one such configuration replaces another, and to which deliberation seems irrelevant, carries with it a *commitment that is exclusive,* we have formed a hybrid notion that is true *neither* to perception *nor* to revolution: Visual apprehension is perhaps plausibly described as intuitive, but it neither demands nor presupposes exclusive commitment. Revolution is plausibly described as demanding such exclusive commitment but it emphatically is not a spontaneous nor intuitive process to which deliberation is, or can be assumed to be, irrelevant.

If we apply such a hybrid notion to the case of scientific theory change, we are led to develop a view that emphasizes the *intuitive and spontaneous* shift of thought and leaves no room for deliberation or interpretation, but we also emphasize that this shift requires exclusive commitment, that it represents a victory for one side in a conflict of loyalties. Such a conception of paradigm change, as *determinative of commitment but itself immune to deliberation,* approximates Kuhn's view of paradigms: they are not altered through interpretation. As he says, "Rather than being an interpreter, the scientist who embraces a new paradigm is like the man wearing inverted lenses" (p. 121). Anomalies and crises "are terminated, not by deliberation and interpretation, but by a relatively sudden and unstructured event like the gestalt switch" (p. 121). Moreover, adopting a new paradigm is not *just* like the vision of the duck for the first time in the duck-rabbit figure: the paradigmatic vision demands commitment, acceptance; it is jealous of other paradigms and seeks to triumph over them. Indeed, its victory enables it to celebrate its triumph by stamping itself into the records, textbooks, and chronicles under its control.

Such a view, insofar as it derives any plausibility from the analogies with vision and with revolution, does not deserve such plausibility. For these analogies, *taken in combination,* form a hybrid which is true to neither. Nor is this simply a matter of the pragmatic *incongruity* of models, a gauche mixing of philosophical metaphors. The mixture perpetuates an old philosophical mistake that consists in deriving values from visions; Plato thus supposed a vision of the form of the good to yield wisdom in life and imagined the role of philosopher-king as that of social authority resting upon superior insight.[9] But values cannot be seen; nor are visions judgments or the premises of judgments. Value judgments, on the other hand, are certainly *judgments;* they make claims and demand acceptance and commitment. Insofar as this is true, they invite discussion, interpretation, and deliberation: they allow of a weighing of evidence and a scrutiny of reasons *pro* and *con*. This process of *deliberation over a claim* cannot be short-circuited by a spontaneous vision or insight. Philosophers have tried hard, to be sure, in the interests of one or another dogmatic and authoritarian conception, to end the deliberative discussion of *pros* and *cons* by authoritative appeals to a favored superior vision. But the attempt can hardly be sustained upon reflection.

Nor can the use of the further analogy with conversion improve the situation. For if converts rest only with the occurrence of mystical experiences, they cannot hope thereby to propound *claims* and *judgments* even though their perspective on life may be fundamentally altered. If, on the other hand, they appeal to evidence and stated considerations, their claims invite debate logically, no matter how resistant they may be to such debate. The matter is further confused by the fact that conversion itself is sometimes defended on the analogy with vision.

How then shall we construe the process of theory-change? If the metaphors of vision and revolution cannot be applied simultaneously, can one or the other be chosen in itself to represent the process, or can both be employed, but not in combination? Let us consider the situation once again and recall also the special features of the analogies in question.

The duck and the rabbit do not make rival claims upon our acceptance. Once you have seen both the duck and the rabbit, you may perhaps be puzzled by various aspects of the reversible figure, but the question as to which is to be exclusively accepted does not arise. There is no choice to be made, no common problem or standard by reference to which such a choice could be made in-

---

[9] Compare K. R. Popper, *The Open Society and its Enemies* (London: Routledge and Kegan Paul, Ltd., 1945, Third edition (revised) 1957), Chapters 7 and 8.

telligible. You accept both "readings" of the puzzle figure as facts and make no decisions of exclusive acceptance. The "readings" themselves make no claims upon your commitment. They may be mutually exclusive as natural occurrences but they are not logically incompatible nor do they generate rival claims and loyalties. Having seen them both, you can rise above them and contain them without logical or moral conflict.

Where theories in science are in opposition, the same cannot be said; the conflict is not simply a matter of inability to hold both theories in view simultaneouly, so to speak, but a conflict of claims demanding resolution. Here there is a choice to be made, a resolution to be sought which will determine acceptances and commitments. You may not be able to *reach* a decision comfortably, or at all, but you cannot tolerantly rise above the situation and adopt a sophisticated nonchalance. In these respects the vision metaphor is clearly inadequate.

Further, the process of resolution is neither relatively sudden nor intuitive nor individual. Kuhn speaks of the "flashes of intuition through which a new paradigm is born" (p. 122). But the birth of a paradigm in the mind does not exhaust the period of its ascendancy and triumph. If "interpretation" does not properly describe the process of *birth,* it certainly describes the processes by which the paradigm is submitted to public scrutiny and runs the gauntlet of debate and criticism which precede its victory. Crises are not, contrary to Kuhn, terminated by gestalt switches (p. 121). The gestalt switch is only the beginning. The new paradigm idea has to be formulated, published, argued, defended, tested, and submitted to examination by colleagues with different preconceptions and with access to various sorts of evidence. In this respect too, the metaphor of the gestalt switch is inadequate, and that of revolution, with its notion of prolonged struggle and ultimate decision, seems superior.

Yet, once we recognize that the *adoption* of a new theory or paradigm is not an instantaneous or individual affair, we may find a certain point to the vision metaphor after all. It is a myth to suppose that paradigms are *units,* that they are simply *accepted,* or simply *rejected* as such, and in a momentary spontaneous process at that. This picture needs complication in many directions. At least one complication involves separation of the *birth* of a paradigm from its *testing* in the public arena.

Now we must not suppose that the *originator* gives it over to others to test; that he himself is, of necessity, fully convinced through the origination itself. On the contrary, the originator's own conviction is, in principle, subject to the vicissitudes of the public process of scientific discussion and evaluation. He originates the idea and there is no way of telling how he does it, nor any way of reducing creativity to rule, but his conviction in his idea depends, at least in part, upon how it fares. When others contribute to testing the originator's idea they must first appreciate it: they seek to determine its content and its logical bearing upon the available evidence.

The process of *grasping or understanding* a theory (whether by the originator or his colleagues) may perhaps plausibly be compared to vision. Indeed *seeing* is a standard way of describing *comprehension*. Intuition is a matter of *seeing the point*. It may indeed be relatively sudden, and like creativity itself, impossible to reduce to routine or mechanical rule. Understanding does not, however, in itself, imply advocacy or commitment, nor is it excluded by rejection of the theory in question.

This is indeed a central point of the scientific attitude. Understanding a theory, our acceptance or denial of it is not thereby prejudged. Our advocacy or rejection itself depends upon the outcome of tests and argumentation; it is not predetermined simply by our comprehension. Conversely, rejecting a theory does not *imply* that we do *not* understand it.

In sum, vision may perhaps appropriately serve as a metaphor for *comprehension* of a paradigm or theory, though not for its testing and acceptance or rejection. The latter involve advocacy and claims to commitment; they involve debates and counterdebates, a period of testing and ultimate ascendancy or decline. In these latter respects, the revolutionary metaphor is appropriate, as earlier argued. But as also remarked earlier, revolutionary debate is not necessarily scientific debate. Advocacy *in itself* is not necessarily fair, logical, or responsive to competent argument or relevant evidence. In these respects, science can only be compared in very limited ways to revolution. The *quality* of scientific deliberations makes for a special and rare form of argumentation.

It demands responsibility to the evidence, openness to argument, commitment to publication, loyalty to logic, and an admission, in principle, that one may turn out to be wrong. These special features of deliberation go far beyond the gestalt switch analogy and they outstrip the revolution metaphor as well. An understanding of science requires appreciation of these special features, a recognition that science itself marks a revolution in the quality of human thought.

Appendix B

In Praise of the Cognitive Emotions*

The mention of cognitive emotions may well evoke emotions of perplexity or incredulity. For cognition and emotion, as everyone knows, are hostile worlds apart. Cognition is sober inspection; it is the scientist's calm apprehension of fact after fact in his relentless pursuit of Truth. Emotion, on the other hand, is commotion —an unruly inner turbulence fatal to such pursuit but finding its own constructive outlets in aesthetic experience and moral or religious commitment.

Strongly entrenched, this opposition of cognition and emotion must nevertheless be challenged for it distorts everything it touches: Mechanizing science, it sentimentalizes art, while portraying ethics and religion as twin swamps of feeling and unreasoned commitment. Education, meanwhile—that is to say, the development of mind and attitudes in the young—is split into two grotesque parts —unfeeling knowledge and mindless arousal. My purpose here is to help overcome the breach by outlining basic aspects of emotion in the cognitive process.

Some misgivings about this purpose will, I hope, be allayed by a preliminary word. My aim, to begin with, is not reductive; I am concerned neither to reduce emotion to cognition nor cognition to emotion—only to show how cognitive functioning employs and incorporates diverse emotional elements—these elements themselves acquiring cognitive significance thereby. I am emphatically

* Originally presented as a special lecture in May 1976 at the 129th annual meeting of the American Psychiatric Association, this paper was also given as a special lecture in 1976 at Brooklyn College. A variant constituted the John Dewey lecture for 1977, sponsored by the John Dewey Society. Versions of the paper were also presented at Harvard University, The University of Notre Dame, Brandeis University, and Long Island University. The paper was published in *Teachers College Record,* vol. 79 (1977), pp. 171–186. Reprinted by permission of the publisher.

not suggesting that cognitions are essentially emotions, or that emotions are, in reality, only cognitions. Nevertheless, I hold that cognition cannot be cleanly sundered from emotion and assigned to science, while emotion is ceded to the arts, ethics, and religion. All these spheres of life involve both fact and feeling; they relate to sense as well as sensibility.

Secondly, though applauding the cognitive import of emotions, I do not propose to surrender intellectual controls to wishful thinking, nor shall I portray the heart as giving special access to a higher truth.[1] Control of wishful thinking is utterly essential in cognition; it operates, however, not through an unfeeling faculty of Reason but through the organization of countervailing critical interests in the process of inquiry. These interests of a critical intellect are, in principle, no less emotive in their bearing than those of wayward wish. The heart, in sum, provides no substitute for critical inquiry; it beats in the service of science as well as of private desire.

Finally, I concede it to be undeniable that certain emotional states may be at odds with sound processes of judgment and decision making. Overpowering agitations may derail the course of reasoning; greed, jealousy, or lust may misdirect it; depression or terror may bring it to a total halt. Conversely, the effect of rational judgment may well be to moderate, even wholly to dissipate, certain emotions by falsifying their factual presuppositions: Anger fades, for example, when it turns out the injury was accidental or caused by someone other than first supposed; fear evaporates when the menacing figure becomes the tree's dancing shadow. It does not follow from these cases, however, that *emotion* as such is uniformly hostile to cognitive endeavors, nor may we properly conclude that *cognition* is, in general, free of emotional engagement. Indeed, emotion without cognition is blind and, as I shall hope particularly to show in the sequel, cognition without emotion is vacuous.

---

[1] For a discussion of this theme in the context of the history of American thought, see Morton White, *Science and Sentiment in America* (New York: Oxford University Press, 1972).

Emotions in the Service of Cognition

Considering now the various roles of emotion in cognition, I divide the field, for convenience, into two main parts, the first having to do with the organization of *emotions generally* in the service of critical inquiry, and the second having to do with *specifically cognitive emotions*. Under the first rubric I shall treat: (a) rational passions,[2] (b) perceptive feelings, and (c) theoretical imagination, and I turn first to the rational passions, that is to say, the emotions undergirding the life of reason.

Rational Passions   The life of reason is one in which cognitive processes are organized in accord with controlling rational ideals and norms. Such organization involves characteristic patterns of thought, action, and evaluation comprising what may be called rational character. It also thus requires suitable emotional dispositions. It demands, for example, a love of truth and a contempt of lying, a concern for accuracy in observation and inference, and a corresponding repugnance of error in logic or fact. It demands revulsion at distortion, disgust at evasion, admiration of theoretical achievement, respect for the considered arguments of others. Failing such demands, we incur rational shame; fulfilling them makes for rational self-respect.

Like moral character, rational character requires that the right acts and judgments be habitual; it also requires that the right emotions be attached to the right acts and judgments.[3] "A rational man," says R. S. Peters, "cannot, without some special explanation,

---

2 On this topic see R. S. Peters, "Reason and Passion," in *Education and the Development of Reason,* eds. R. F. Dearden, P. H. Hirst, and R. S. Peters (London, U.K.: Routledge and Kegan Paul, 1972), especially the section on the rational passions, pp. 225–27. See also John Rawls, *A Theory of Justice* (Cambridge, Mass.: Harvard University Press, 1971), especially sections 67, 73–75; P. Foot, "Moral Beliefs," *Proceedings of the Aristotelian Society,* 59, 1958–1959; and B. A. O. Williams, "Morality and the Emotions," in *Problems of the Self,* B. A. O. Williams (Cambridge, U.K.: Cambridge University Press, 1973). A significant recent book, dealing with a wide range of related topics, is Robert C. Solomon, *The Passions* (New York: Doubleday, Anchor Press, 1976).

3 Cp. Aristotle, *Nicomachean Ethics,* Book II, 3.

slap his sides and roar with laughter or shrug his shoulders with indifference if he is told that what he says is irrelevant, that his thinking is confused and inconsistent or that it flies in the face of the evidence."[4] The suitable deployment in conduct of emotional dispositions such as love and hate, contempt and disgust, shame and self-esteem, respect and admiration indeed defines what is meant, quite generally, by the internalization of ideals and principles in character. The wonder is not that *rational* character is thus related to the emotions but that anyone should ever have supposed it to be an exception to the general rule.

Rational character constitutes an intellectual conscience; it monitors and curbs evasions and distortions; it combats inconsistency, unfairness to the facts, and wishful thinking. In thus exercising control over undesirable impulses, it works for a balance in thought, an epistemic justice, which requires its own special renunciations and develops a characteristic cognitive discipline. There is, however, no question here of the control of impulses through a "bloodless reason,"[5] as control is exercised through the structuring of emotions themselves. Rationality, as John Dewey put it,

> is not a force to evoke against impulse and habit. It is the attainment of a working harmony among diverse desires. . . . The elaborate systems of science are born not of reason but of impulses at first slight and flickering; impulses to handle, move about, to hunt, to uncover, to mix things separated and divide things combined, to talk and to listen. Method is their effectual organization into continuous dispositions of inquiry, development and testing. It occurs after these acts and because of their consequences. Reason, the rational attitude, is the resulting disposition. . . . The man who would intelligently cultivate intelligence will widen, not narrow, his life of strong impulses while aiming at their happy coincidence in operation.[6]

This coincidence, I emphasize, requires appropriate organization of feelings and sentiments in the interests of intelligent control.

[4] Peters, "Reason and Passion," p. 226.

[5] John Dewey, *Human Nature and Conduct* (New York: Henry Holt, 1922, 1930), p. 196.

[6] *Ibid.*

Perceptive Feelings   Having seen the role of emotions in the internalization of rational norms, let us consider now their employment in perception. For they are not only interwoven with our cognitive ideals and evaluative principles; they are also intimately tied to our vision of the external world. Indeed they help to construct that vision and to define the critical features of that world.

These critical features—however specified—are the objects of our evaluative attitudes, the foci of our appraisals of the environment. Our habits and judgments are keyed in to these appraisals; we define ourselves and orient our action in the light of our situation as appraised. Characteristic orientations are associated with distinctive emotional dispositions, and both involve seeing the environment in a certain light: Is it, for example, beneficial or harmful, promising or threatening, fulfilling or thwarting?[7] The subtle and intricate web relating adult feeling and orientation to adult perception of the environment is a product of evolutionary development, to be sure, but also of the special circumstances of individual biography. Acquiring human significance through biographical linkage with critical features of the environment, our feelings come indeed to *signify*—to serve as available cues for interpreting the situation.

Fear of a particular person, for example, presupposes that that person is regarded as dangerous—danger being a critical feature of the environment calling for a special orientation in response. There need, however, be no *independent* evidence, in every case, of the threat we sense: The characteristic feeling that has become associated for us with past dangers itself serves us as a cue. Interpreting that feeling *as* fear, we at once characterize our own state and ascribe danger to the environment. Indeed, we may thence proceed to an explicit attribution of danger, prompted by cues of

---

[7] Related points are discussed in Peters, "Reason and Passion"; R. S. Peters, "The Education of the Emotions," Dearden, Hirst, and Peters, *Education and the Development of Reason;* G. Pitcher, "Emotion," Dearden, Hirst, and Peters, *Education and the Development of Reason;* and R. W. Hepburn, "The Arts and the Education of Feeling and Emotion," Dearden, Hirst, and Peters, *Education and the Development of Reason.* See also the article by W. P. Alston, "Emotion and Feeling," in *The Encyclopedia of Philosophy* (New York: Macmillan, 1967), vol. 2, pp. 479–86.

feeling. Pursuing a more abstract direction in forming our cognitive concepts, we may, further, come to describe a certain situation as *terrifying,* ascribing to it, *independently of our own state,* the capacity to arouse fear. Thus employing the emotions as parameters, we gain enormous new powers of fundamental description, while abstracting from actual conditions of feeling.

The notion that aesthetic experience, for example, is peculiarly and purely a matter of emotion ignores such manifold connections of feeling and fact—both fact as *embodied* in the art work and fact as *represented* therein. Relative to the latter, H. D. Aiken writes,

> Just as in ordinary circumstances an emotional response is the product of a perceived situation which is apprehended by the individual as promising or threatening, so the expressiveness of an imaginative work arises, at least in part, from the fact that it provides a dramatic representation of an action of which the evoked emotion is the expressive counterpart. And such a representation must be understood as such if the expressive values of the work are to become actual; without it such emotion as the observer might experience would have no ground, and if, by a miracle, it could be sustained, it would still remain the private, dumb, inexpressive importation of the observer himself. As such it would be nothing more than an accidental, adventitious subjective coloring which, having no artistic basis in the thing perceived, would be devoid of aesthetic relevance to it. Aesthetically relevant emotion in art is something which is expressed to us by the action or gesture of the work itself; it is something aroused and sustained by the work as an object for contemplation, and it is found there as a projected quality of the action.[8]

That the emotion is thus tied to a representational understanding of the work of art does not imply, however, that this understanding must be antecedently fashioned, in complete isolation from the feelings. This point must be especially emphasized since the familiar notion of the work of art as "an object for contemplation" may carry contrary, and therefore misleading, connotations. In fact, I believe, the very feelings through which we respond to the content of a work serve us also in interpreting this content.

[8] Henry David Aiken, "Some Notes Concerning the Aesthetic and the Cognitive," *Journal of Aesthetics and Art Criticism,* 13(1955), 390–91.

Reading our feelings and reading the work are, in general, virtually inseparable processes.

The cognitive role of the emotions in aesthetic contexts has been emphasized by Nelson Goodman in a recent discussion. He writes,

> The work of art is apprehended through the feelings as well as through the senses. Emotional numbness disables here as definitely if not as completely as blindness or deafness. Nor are the feelings used exclusively for exploring the emotional content of a work. To some extent, we may feel how a painting looks as we may see how it feels. The actor or dancer—or the spectator—sometimes notes and remembers the feeling of a movement rather than its pattern, insofar as the two can be distinguished at all. Emotion in aesthetic experience is a means of discerning what properties a work has and expresses.[9]

The general point is, of course, not limited to the aesthetic realm since, as I have earlier emphasized, the emotions intimately mesh with all critical appraisals of the environment: The flow of feeling thus provides us with a continuous stream of cues significant for orientation to our changing contexts. Indeed, as Goodman remarks,

> In daily life, classification of things by feeling is often more vital than classification by other properties: we are likely to be better off if we are skilled in fearing, wanting, braving, or distrusting the right things, animate or inanimate, than if we perceive only their shapes, sizes, weights, etc. And the importance of discernment by feeling does not vanish when the motivation becomes theoretic rather than practical. . . . Indeed, in any science, while the requisite objectivity forbids wishful thinking, prejudicial reading of evidence, rejection of unwanted results, avoidance of ominous lines of inquiry, it does not forbid use of feeling in exploration and discovery, the impetus of inspiration and curiosity, or the cues given by excitement over intriguing problems and promising hypotheses.[10]

Theoretical Imagination   Mention of the context of theory brings us to the third role of emotions in the service of cognition, that of stimulus to the scientific imagination. This role is virtually

---

[9] Nelson Goodman, *Languages of Art* (Indianapolis, Ind.: Hackett Publishing Co., 1968, 1976), p. 248.

[10] *Ibid.*, p. 251.

annihilated by the stereotyped emotion-cognition dichotomy. For this dichotomy assigns all feeling and flair, all fantasy and fun, to the arts and humanities, conceiving the sciences as grim and humorless grind. The method of science is miserly caution—to gather the facts and guard the hoard. Imagination is a seductive distraction —a hindrance to serious scientific business.

This doctrine is, in fact, the death of theory. Theory is not reducible to mere fact-gathering, and theoretical creation is beyond the reach of any mechanical routine. Science controls theory by credibility, logic, and simplicity; it does not provide rules for the creation of theoretical ideas. Scientific objectivity demands allegiance to fair controls over theory, but fair controls cannot substitute for ideas. "All our thinking," said Albert Einstein, "is of this nature of a free play with concepts; the justification for this play lying in the measure of survey over the experience of the senses which we are able to achieve with its aid."[11]

The ideal theorist, loyal to the demands of rational character and the institutions of scientific objectivity, is not therefore passionless and prim. Theoretical inventiveness requires not caution but boldness, verve, speculative daring. Imagination is no hindrance but the very life of theory, without which there *is* no science.

Now the emotions relate to imaginative theorizing in a variety of ways. The emotional life, to begin with, is a rich source of substantive ideas. Drawing from the obscure wellsprings of this life, the mind's free play casts up novel patterns and images, exotic figures and analogies that, in an investigative context, may serve to place old facts in a new light. The dream of the nineteenth-century chemist F. A. Kekulé will provide a striking illustration. He had been trying for a long time to find a structural formula for the benzene molecule. Dozing in front of his fireplace one evening in 1865, he seemed, as he looked into the flames, "to see atoms dancing in snakelike arrays. Suddenly, one of the snakes formed

---

[11] Albert Einstein, "Autobiographical Notes," tr. Paul Arthur Schilpp, in *Albert Einstein: Philosopher-Scientist,* ed. P. A. Schilpp (New York: Tudor Publishing, 1949), p. 7. (Now published by the Open Court Publishing Company, La Salle, Illinois.) The passage is quoted in a discussion of these and related points in chapter 4 above.

a ring by seizing hold of its own tail and then whirled mockingly before him. Kekulé awoke in a flash: he had hit upon the now famous and familiar idea of representing the molecular structure of benzene by a hexagonal ring. He spent the rest of the night working out the consequences of this hypothesis."[12]

The emotions serve not merely as a *source* of imaginative patterns; they fulfill also a *selective* function, facilitating choice among these patterns, defining their salient features, focusing attention accordingly. The patterns developed in imagination, that is, carry their own emotive values; these values guide selection and emphasis. They help imagined patterns to structure the phenomena, highlighting factual features of interest to further inquiry. "Passions," as Michael Polanyi has said, "charge objects with emotions, making them repulsive or attractive; . . . Only a tiny fraction of all knowable facts are of interest to scientists, and scientific passion serves . . . as a guide in the assessment of what is of higher and what of lesser interest."[13]

Finally, the emotions play a directive role in the process of *applying* the fruits of imagination to the solution of problems. The course of problem-solving, as has already been intimated, is continually monitored by the theorist's cues of feeling, his sense of excitement or anticipation, his elation or suspicion or gloom. Moreover, imagined objects encountered in thought by the problem-solver affect his deliberation emotively, as real objects do, and influence his decisions in analogous ways. "In thought as well as in overt action," says Dewey, "the objects experienced in following out a course of action attract, repel, satisfy, annoy, promote and retard. Thus deliberation proceeds."[14] There is, no doubt, much yet to be learned about the interaction of emotions and imagination in all the ways I have sketched, and in others as well. It should, however, even now, be evident that creation is fed by the

---

[12] Carl G. Hempel, *Philosophy of Natural Science* (Englewood Cliffs, N.J.: Prentice-Hall, 1966), p. 16.

[13] Michael Polanyi, *Personal Knowledge* (New York: Harper & Row, 1958, 1962), pp. 134–35.

[14] Dewey, *Human Nature and Conduct*, p. 192.

emotional life in the sphere of science no less than in the spheres of poetry and the arts.

Cognitive Emotions

We have, until now, concerned ourselves with the organization of emotions generally in the service of cognition. I want now to deal with two emotions that are, in a sense to be explained, *specifically cognitive* in their bearing—the *joy of verification* and the *feeling of surprise.*

In what sense do I speak of an emotion as specifically cognitive? Consider first the notion of *moral emotions,* conceived as those resting upon suppositions of a moral sort: Thus, indignation, for example, rests upon the supposition of a moral grievance—a piece of injustice, and remorse presumes that one has in fact done something wrong. If the relevant moral suppositions are false or lack evidential foundation, the respective emotions may be thought unreasonable, but if these suppositions are not made at all, that is to say, if the suppositions do not exist, the emotions in question can hardly, in normal circumstances, be said to have occurred. Now I propose, analogously, to consider an emotion specifically *cognitive* if it rests upon a supposition of a cognitive sort—that is to say, a supposition relating to the content of the subject's cognitions (beliefs, predictions, expectations) and, in cases of special interest to us, bearing upon their epistemological status.

It is important to avoid misunderstanding of the terminology I have chosen here. It is indeed true that *all* suppositions may be considered cognitive in a broad sense, inasmuch as they make factual claims expressible in propositional form; moreover, emotions *generally,* as I have maintained, presuppose the existence of such claims concerning critical features of the environment. However, when I characterize an emotion as *specifically cognitive,* I mean more than this. In particular, I mean not simply that it presupposes the existence of a factual claim but that the claim in question specifically concerns the nature of the subject's cognitions (and, in cases of interest, is epistemologically relevant to them). A cognitive emotion, I should further emphasize, is thus decidedly an *emotion,* but an emotion of a certain *kind,* specifiable by its cognitive reference as just explained.[15]

The Joy of Verification In his well-known paper of 1934 on "The Foundation of Knowledge,"[16] Moritz Schlick provides an example of such a cognitive emotion in outlining his theory of science, giving primary place in his theory to the joy that accompanies the fulfillment of an expectation. Cognition, in Schlick's view, has, from the earliest times, always been predictive but the value of reliable prediction lay originally in its practical service to life. "Now in science," he writes, ". . . cognition . . . is not sought because of its utility. With the confirmation of prediction the scientific goal is achieved: the joy in cognition is the joy of verification, the triumphant feeling of having guessed correctly."[17] Such moments of joy are, in Schlick's opinion, of central importance in understanding scientific purpose. "They do not in any way," he says, "lie at the base of science; but like a flame, cognition, as it were, licks out to them, reaching each but for a moment and then at once consuming it. And newly fed and strengthened, it flames onward to the next. These moments of fulfillment and combustion are what is essential. All the light of knowledge comes from them."[18]

Now one need not agree with Schlick's general view of science in order to acknowledge that the satisfaction of a theoretical forecast may indeed occasion joy. Nor is it required that we concur with the extravagant suggestion that *all* predictive success brings elation. It may, for example, be countered that routine successes based on theory frequently, perhaps typically, go unnoticed, while

---

[15] For discussion helpful in clarifying certain points in this section, I am grateful to Professors Eli Hirsch and Jonas Soltis.

[16] Moritz Schlick, "Über das Fundament der Erkenntnis," *Erkenntnis,* vol. 4, 1934, trans. as David Rynin, "The Foundation of Knowledge," in *Logical Positivism,* ed. A. J. Ayer (New York: Free Press, 1959).

[17] Schlick, Über das Fundament der Erkenntnis," in Ayer, *Logical Positivism,* pp. 222–23. There is a general discussion of Schlick's paper in chapter 5 above.

[18] Schlick, "Über das Fundament der Erkenntnis," in Ayer, *Logical Positivism,* p. 227.

soberly predicted events may be so dreadful as to occasion not joy but sorrow or despair. Nevertheless, we can agree that the fulfillment of a prediction may indeed crown an investigative achievement in science, producing in its wake what Schlick calls a "triumphant feeling of having guessed correctly."

This joyful feeling I consider a cognitive emotion, because it rests on a supposition (with epistemological relevance) as to the content of the guess in question: It presumes that what has happened is what had, in fact, been predicted. Without such presumption, this joy of verification cannot be said to occur. Whether the presumption is true, or is based on adequate grounds, is another story. Certainly one assumes that the emotion in question may be criticized as unreasonable if it can be shown that what has in fact happened had not, in fact, been predicted.

Can such a criticism, however, actually be entertained as a matter of psychological fact? Is it not, rather, true that our expectations are so powerful as consistently to warp our observations to fit? A whole library of psychological writings testifies to the powerful tendency of expectation to create its own confirmations in experience. The general theme of this testimony may be indicated by the role of normal cues in perception: The perceptual identification of objects "proceeds on the basis of cues normally sufficient to select the objects in question; when these cues fail in fact, we tend anyway to see them as having succeeded."[19] Bruner, Goodnow, and Austin comment on this point as follows:

> If [a bird] has wings and feathers, the bill and legs are highly predictable. In coding or categorizing the environment, one builds up an expectancy of all of these features being present together. It is this unitary conception that has the configurational or Gestalt property of "birdness". Indeed, once a configuration has been established and the object is being identified in terms of configurational attributes, the perceiver will tend to "rectify" or "normalize" any of the original defining attributes that deviate from expectancy. Missing attributes are "filled in" . . . , reversals righted . . . , colors assimilated to expectancy.[20]

[19] P. 30 above.

[20] Jerome S. Bruner, J. J. Goodnow, and G. A. Austin, *A Study of Thinking* (New York: John Wiley, 1966), p. 47.

Some philosophers have further maintained that scientific observation itself is *systematically* theory-laden—presupposing the very theories it is naively thought to test.[21] If indeed we are, as suggested, so blinded by our own theoretical beliefs as to be incapable of acknowledging anything that might contradict them, we can hardly take the joy of verification to represent a cognitive triumph of science. Rather, we must count it an unearned and deluded joy, resulting not from a happy match between theory and experience but solely from our desperate rigging of experience to make it fit.

This conclusion, as I have elsewhere argued, seems to me too extreme for the facts.[22] It is undeniable that our beliefs greatly influence our perceptions, but neither psychology nor philosophy offers any proof of a preestablished harmony between what we believe and what we see. Expectations have the function of orienting us selectively toward the future, but this function does not require that they blind us to the unforeseen. Indeed, the presumption of mismatch between experience and expectation underlies another cognitive emotion: *surprise.* The existence of this emotion testifies that we are not, in principle, beyond acknowledging the predictive failures of our own theories, that we are not debarred by nature from capitalizing upon such failures in order to learn from experience. The genius of science is, in fact, to institutionalize such learning by wedding the free theoretical imagination to the rigorous probing for predictive failures.

The Significance of Surprise Surprise is a cognitive emotion, resting on the (epistemologically relevant) supposition that what has happened conflicts with prior expectation. Without such presumption, surprise cannot be supposed to occur, although the truth of the presumption may, of course, be questioned in particular cases. Surprise must, in any event, not be confused with mere novelty. A novel—that is to say, a hitherto unencountered—contingency may well be anticipated in thought, while a familiar phenomenon, juxtaposed with available theory, may profoundly surprise.

---

[21] See N. R. Hanson, *Patterns of Discovery* (Cambridge, U.K.: Cambridge University Press, 1958), pp. 18–19, and elsewhere. For a general discussion see chapters 1 and 2 above.

[22] *Ibid.*, chapter 2.

Thwarting expectation, the surprising element may indeed provoke the revision of theory, even the reorganization of categories, thus *producing* novelty as a result. It is itself, however, never a mere matter of novelty, but always of conflict with prior belief. The concept of *unexpectedness,* it should be noted, is too weak to make this critical emphasis, for it covers both the case of a feature that has simply not been anticipated, and that of a feature that has been positively ruled out by anticipation.[23]

To the extent that we are capable of surprise, the possibility that our expectations are wrong is alive for us and thus our joy in verification, if it occurs, is not utterly deluded. Receptive to surprise, we are capable of learning from experience—capable, that is, of acknowledging the inadequacies of our initial beliefs, and recognizing the need for their improvement. It is thus that the testing of theories, no less than their generation, calls upon appropriate emotional dispositions.

Receptivity to surprise involves, however, a certain vulnerability; it means accepting the risk of a possibly painful unsettlement of one's beliefs, with the attendant need to rework one's expectations and redirect one's conduct. To be sure, where the relevant beliefs are weakly held, or relatively segregated, or of peripheral significance for one's basic orientation—or where the required alterations are likely to be readily effected—the risk may be easily borne, even by the cautious. Surprise may, in such circumstances, not in fact distress but amuse—even enchant, as will be evident from even brief reflection on the role of surprise in humor, in music, in litera-

---

[23] On these points, there is disagreement among previous writers. For informative historical, as well as other material see M. M. Desai, "Surprise: A Historical and Experimental Study," *British Journal of Psychology Monograph Supplements,* No. 22, 1939; D. E. Berlyne, "Emotional Aspects of Learning," *Annual Review of Psychology,* 15 (1964), 115–42; and W. R. Charlesworth, "The Role of Surprise in Cognitive Development," in *Studies in Cognitive Development,* D. Elkind and J. H. Flavell (New York: Oxford University Press, 1969). Although I agree with various points in these psychological papers (e.g., Charlesworth, pp. 270, 276), they tend to focus on individual behavior in a relatively local situation, whereas I tend to link surprise with failed prediction in the context of discussions in philosophy of science.

ture, and generally in the arts. Moreover, there are, in all realms of life, pleasant surprises, where the value of the unexpected event, or even of the unsettlement itself, outweighs the stress of disorientation and the concomitant costs of revision in belief.

One cannot, however, reasonably count on all—even most—surprises in life to be thus amusing or pleasant; it must be conceded that a general openness to surprise involves a real risk of epistemic distress. This risk may to varying degrees become palatable, even exciting; certainly accepting it is one of the normal requirements of rational character. Yet it *is* a risk of possibly painful disorientation, and it requires emotional strength to face and to master. To commit oneself to learning from experience is, in short, a significant attitude—supported by mature reflection, to be sure—but exacting a price in return for the prospect of improvement in one's system of beliefs.

Three alternative attitudes promise an avoidance of the price by erecting wholesale defenses against surprise. Since surprise presumes prior expectation, a defense may be sought, to begin with, in the *rejection* of all expectation—in effect, the denial of all belief. This is the attitude of the radical skeptic, who hopes to make himself immune to surprise by any contingency through renouncing all anticipations to the contrary, that is to say, all anticipations without exception. Of whatever happens, he says, in effect, "It doesn't surprise me since I never expected it not to happen!" A second—apparently opposite—attitude is that of utter credulity or gullibility—the *acceptance* of all beliefs or expectations without distinction. Here the formula in response to every contingency is "I'm not surprised! I expected that too!" Both radical skepticism and radical credulity are, however, alike forms of epistemic apathy: To reject all expectations is to be indifferent to each, while to accept all as equally good is in actuality to choose none, having no reason to expect anything at all rather than something else. It is no wonder that these seemingly contrary attitudes have so often been remarked to be psychologically akin, and together opposed to the selective hypothesis-formation characteristic of scientific thought. "Complete doubt," as Peirce noted, is "a mere self-deception" and no one who follows the method of radical skepticism "will ever be satisfied until he has formally recovered all those beliefs which in form he has given up."[24]

Moreover, each of these two alternative attitudes exacts its own heavy price. Neither can in fact be realized as a genuine option over a significant area of conduct. The skeptic, despite himself, forms positive expectations in executing his actions, while the radically credulous person, generally hospitable to inconsistencies, perforce rules out certain contingencies in carrying through the activities of daily life. Only in a local and intermittent way can these attitudes be attempted. They are perhaps more accurately described as poses or pretenses, the effect of which is, however, perfectly real—to aid the denial of responsibility for one's beliefs and so to block the possibility of their improvement through the educative medium of surprise.

The third attitude promising a defense against surprise is that of dogmatism. Unlike the radical skeptic and the total believer, the dogmatist is perfectly firm about the beliefs he espouses and the beliefs he rejects. He blocks surprise not by disclaiming responsibility for his doctrines, but rather by denying all experience that purports to contradict them. He not only avoids the systematic testing of his beliefs; he closes off the very possibility of recognizing negative evidence, by early and stout denial of its existence. Theory-laden to the point of blindness, his observations are predictably positive, the joy he takes in verification thus unearned and hollow. Dogmatism is also a difficult attitude to maintain if only because (as Peirce saw)[25] it is impossible to filter all negative indications in advance by a systematic method. Yet it can be carried a long way, preventing the acknowledgment of surprise, and, hence, the application of new and surprising experience to the improvement

[24] C. S. Peirce, "Some Consequences of Four Incapacities," in *Collected Papers of Charles Sanders Peirce,* ed. Charles Hartshorne and Paul Weiss, vol. 5 (Cambridge: Harvard University Press, 1934), 5:264–65. See also Israel Scheffler, *Four Pragmatists* (London, U.K.: Routledge & Kegan Paul, 1974), pp. 52–53, 69–70.

[25] C. S. Peirce, "The Fixation of Belief," Hartshorne and Weiss, *Collected Papers of Charles Sanders Peirce,* 5:382. See also a general discussion in Scheffler, *Four Pragmatists,* p. 60ff.

of initial beliefs and orientations. Dogmatism, no less than skepticism and gullibility, conflicts with the effort at such improvement. To accept this effort, with its associated vulnerability to the unsettlement of surprise, is to choose a distinctive emotional as well as cognitive path.

But how, it may be asked, is receptivity to surprise possible? Surprise is, after all, unsettling; it risks the distress of disorientation and the potential pain of relearning. In similar vein, Schlick contrasts the joy of verification with the disappointment of falsification[26]—the disappointment following upon the violation of beliefs in which we had put our trust. How can one counsel receptivity to surprise: Is this not an impossibly mixed emotion, like elation at despair, or happiness at depression?

We must, first of all, reject the suggestion that surprise is always painful. If Schlick's notion of uniformly joyful verification is to be rejected as extravagant, in line with our earlier remarks, his parallel notion of uniformly disappointing falsification must equally be criticized. Some falsifications are, as we have previously intimated, delightful, some disruptions of expectation pleasantly exciting, some occasions of relearning fraught with engaging challenge. Schlick, I suggest, confuses expectation with hope, but the two are clearly separable, and only the former is necessary for surprise.

Yet if surprise is not always painful, surely it sometimes is: It must therefore be conceded, at the very least, to be *uncertain* in its quality. The question then recurs in a new version: How can one counsel receptivity to uncertainty? Here, however, an immediate reply is forthcoming. The original version of the question raised the issue of impossibly mixed emotions, whereas the present version no longer does so. For uncertainty is not an emotion; it is rather a prospect or condition, while the *feeling* of uncertainty mixes readily with receptive and aversive attitudes. Uncertainty is indeed consistently faced in varying ways: Some persons tend to shrink from, while others tend to welcome the prospect. The receptivity to surprise that is implicated in the capacity to learn from

[26] Schlick, "Über das Fundament der Erkenntnis," in Ayer, *Logical Positivism*, p. 223.

experience is, in any case, perfectly coherent in its emotional composition.

Moreover, *receptivity* to surprise is not to be confused with elation or happiness. It is rather the capability of acknowledging surprise than joy in its occurrence that is here in point. Analogously, to acknowledge one's grief does not entail being elated by it. Acknowledgment itself is a possible and a significant attitude, opening the way beyond the acknowledged circumstance.

Yet receptivity is, of course, not enough to characterize the testing phase of inquiry. How we cope with surprise, once it is acknowledged, is of critical importance. Surprise may be dissipated and evaporate into lethargy. It may culminate in confusion or panic. It may be swiftly overcome by a redoubled dogmatism. Or it may be transformed into wonder or curiosity, and so become an educative occasion. Curiosity replaces the impact of surprise with the demand for explanation;[27] it turns confusion into question. To answer the question is to reconstruct initial beliefs so that they may consistently incorporate what had earlier been unassimilable. It is to provide an improved framework of premises by which the surprising event might have been anticipated and for which parallel events will no longer surprise.

Critical inquiry in pursuit of explanation is a constructive outcome of surprise, transforming initial disorientation into motivated search. There is, as we have seen, no mechanical routine that guarantees success in the search for explanatory theory. Yet, an emotional value of such search is to offer mature consolation for the stress of surprise and the renunciation of inadequate beliefs.[28]

---

[27] On explanation generally, see Israel Scheffler, *Anatomy of Inquiry* (New York: Alfred A. Knopf, 1963, Part I; also Hackett Publishing Co., 1981). I here use the term in a very broad sense.

[28] Interesting discussion of this sort of point and related psychological issues is contained in Fay H. Sawyier, "About Surprise" (Paper read to the annual meeting of the Western Division of the American Philosophical Association, St. Louis, Missouri, 1974). For a discussion of the pedagogical use of surprise in the teaching of mathematics see Stephen I. Brown, "Rationality, Irrationality and Surprise," *Mathematics Teaching: The Bulletin of The Association of Teachers of Mathematics,* No. 55, Summer 1971.

Achieving superordinate status in the economy of science, the value of inquiry becomes, indeed, autonomous, pressing new explanations deliberately into situations of risk, testing their vulnerability in novel ways, exposing their implicit predictions systematically to the chance of new surprise.

The constructive conquest of surprise is registered in the achievement of new explanatory structures, while cognitive application of these structures provokes surprise once more. Surprise is vanquished by theory, and theory is, in turn, overcome by surprise. Cognition is thus two-sided and has its own rhythm; it stabilizes and coordinates; it also unsettles and divides. It is responsible for shaping our patterned orientations to the future, but it must also be responsive to the insistent need to learn from the future. Establishing habits, it must stand ready to break them. Unlearning old ways of thought, it must also power the quest for new, and greater, expectations.[29] These stringent demands upon our cognitive processes also constitute stringent demands upon our emotional capacities. The growth of cognition is thus, in fact, inseparable from the education of the emotions.

[29] On related points see the papers by H. Gardner, M. W. Wartofsky, and N. Goodman in *The Monist,* 58 (1974), 319–42.

Index